Selected Titles in This Series

738 **Yasuro Gon,** Generalized Whittaker functions on $SU(2,2)$ with respect to the Siegel parabolic subgroup, 2002

737 **Arjen Doelman, Robert A. Gardner, and Tasso J. Kaper,** A stability index analysis of 1-D patterns of the Gray-Scott model, 2002

736 **Wojciech Chachólski and Jérôme Scherer,** Homotopy theory of diagrams, 2002

735 **Martina Brück, Xi Du, Joonsang Park, and Chuu-Lian Terng,** The submanifold geometries associated to Grassmannian systems, 2002

734 **Michel Van den Bergh,** Blowing up of non-commutative smooth surfaces, 2001

733 **Milé Krajčevski,** Tilings of the plane, hyperbolic groups and small cancellation conditions, 2001

732 **Jan O. Kleppe, Juan C. Migliore, Rosa Miró-Roig, Uwe Nagel, and Chris Peterson,** Gorenstein liaison, complete intersection liaison invariants and unobstructedness, 2001

731 **Jesús Bastero, Mario Milman, and Francisco J. Ruiz,** On the connection between weighted norm inequalities, commutators and real interpolation, 2001

730 **Suhyoung Choi,** The decomposition and classification of radiant affine 3-manifolds, 2001

729 **Michael Grosser, Eva Farkas, Michael Kunzinger, and Roland Steinbauer,** On the foundations of nonlinear generalized functions I and II, 2001

728 **Laura Smithies,** Equivariant analytic localization of group representations, 2001

727 **Anthony D. Blaom,** A geometric setting for Hamiltonian perturbation theory, 2001

726 **Victor L. Shapiro,** Singular quasilinearity and higher eigenvalues, 2001

725 **Jean-Pierre Rosay and Edgar Lee Stout,** Strong boundary values, analytic functionals, and nonlinear Paley-Wiener theory, 2001

724 **Lisa Carbone,** Non-uniform lattices on uniform trees, 2001

723 **Deborah M. King and John B. Strantzen,** Maximum entropy of cycles of even period, 2001

722 **Hernán Cendra, Jerrold E. Marsden, and Tudor S. Ratiu,** Lagrangian reduction by stages, 2001

721 **Ingrid C. Bauer,** Surfaces with $K^2 = 7$ and $p_g = 4$, 2001

720 **Palle E. T. Jorgensen,** Ruelle operators: Functions which are harmonic with respect to a transfer operator, 2001

719 **Steve Hofmann and John L. Lewis,** The Dirichlet problem for parabolic operators with singular drift terms, 2001

718 **Bernhard Lani-Wayda,** Wandering solutions of delay equations with sine-like feedback, 2001

717 **Ron Brown,** Frobenius groups and classical maximal orders, 2001

716 **John H. Palmieri,** Stable homotopy over the Steenrod algebra, 2001

715 **W. N. Everitt and L. Markus,** Multi-interval linear ordinary boundary value problems and complex symplectic algebra, 2001

714 **Earl Berkson, Jean Bourgain, and Aleksander Pełczynski,** Canonical Sobolev projections of weak type $(1,1)$, 2001

713 **Dorina Mitrea, Marius Mitrea, and Michael Taylor,** Layer potentials, the Hodge Laplacian, and global boundary problems in nonsmooth Riemannian manifolds, 2001

712 **Raúl E. Curto and Woo Young Lee,** Joint hyponormality of Toeplitz pairs, 2001

711 **V. G. Kac, C. Martinez, and E. Zelmanov,** Graded simple Jordan superalgebras of growth one, 2001

710 **Brian Marcus and Selim Tuncel,** Resolving Markov chains onto Bernoulli shifts via positive polynomials, 2001

709 **B. V. Rajarama Bhat,** Cocylces of CCR flows, 2001

(*Continued in the back of this publication*)

Generalized Whittaker Functions on $SU(2, 2)$ with Respect to the Siegel Parabolic Subgroup

Memoirs of the American Mathematical Society

Number 738

Generalized Whittaker Functions on $SU(2,2)$ with Respect to the Siegel Parabolic Subgroup

Yasuro Gon

January 2002 • Volume 155 • Number 738 (fourth of 5 numbers) • ISSN 0065-9266

American Mathematical Society
Providence, Rhode Island

2000 *Mathematics Subject Classification*.
Primary 11F70; Secondary 22E45.

Library of Congress Cataloging-in-Publication Data

Gon, Yasuro.
Generalized Whittaker functions on SU(2,2) with respect to the Siegel parabolic subgroup / Yasuro Gon.
p. cm. — (Memoirs of the American Mathematical Society, ISSN 0065-9266 ; no. 738)
"January 2002."
"Volume 155, number 738 (fourth of 5 numbers)."
Includes bibliographical references.
ISBN 0-8218-2763-4 (alk. paper)
1. Coulomb functions. 2. Forms, Modular. 3. Representations of groups. I. Title. II. Series.

QA3.A57 no. 738
[QA353.H9]
510 s—dc21
[512′.73] 2001045784

Memoirs of the American Mathematical Society

This journal is devoted entirely to research in pure and applied mathematics.

Subscription information. The 2002 subscription begins with volume 155 and consists of six mailings, each containing one or more numbers. Subscription prices for 2002 are $524 list, $419 institutional member. A late charge of 10% of the subscription price will be imposed on orders received from nonmembers after January 1 of the subscription year. Subscribers outside the United States and India must pay a postage surcharge of $31; subscribers in India must pay a postage surcharge of $43. Expedited delivery to destinations in North America $35; elsewhere $130. Each number may be ordered separately; *please specify number* when ordering an individual number. For prices and titles of recently released numbers, see the New Publications sections of the *Notices of the American Mathematical Society*.

Back number information. For back issues see the *AMS Catalog of Publications*.

Subscriptions and orders should be addressed to the American Mathematical Society, P. O. Box 845904, Boston, MA 02284-5904. *All orders must be accompanied by payment.* Other correspondence should be addressed to Box 6248, Providence, RI 02940-6248.

Memoirs of the American Mathematical Society is published bimonthly (each volume consisting usually of more than one number) by the American Mathematical Society at 201 Charles Street, Providence, RI 02904-2294. Periodicals postage paid at Providence, RI. Postmaster: Send address changes to Memoirs, American Mathematical Society, P. O. Box 6248, Providence, RI 02940-6248.

This publication is indexed in *Science Citation Index*®, *SciSearch*®, *Research Alert*®, *CompuMath Citation Index*®, *Current Contents*®/*Physical, Chemical & Earth Sciences*.
Printed in the United States of America.

♾ The paper used in this book is acid-free and falls within the guidelines established to ensure permanence and durability.
Visit the AMS home page at URL: `http://www.ams.org/`

10 9 8 7 6 5 4 3 2 1 07 06 05 04 03 02

Contents

Abstract

We obtain an explicit formula for generalized Whittaker functions and multiplicity one theorem for all discrete series representations of $SU(2,2)$.

Received by the editor June 14, 1999 .

1991 *Mathematics Subject Classification.* Primary 11F70; Secondary 22E45.

CHAPTER 1

Introduction

1. Introduction

In this article we study generalized Whittaker functions on $G = SU(2,2)$, the special unitary group of signature $(2+, 2-)$, with respect to the Siegel parabolic subgroup P_S. The subgroup P_S is a maximal parabolic subgroup with abelian unipotent radical N_S. Our concern is to extend the classical theory of Fourier expansion for holomorphic modular forms on G to that for non-holomorphic modular forms. The reason why we investigate this group is because it is of real rank two and has many new features that never exists for smaller groups.

Let π be an admissible representation of G and ξ be a unitary character of N_S. Then a fundamental problem for the theory of Fourier expansion of automorphic forms for π is to investigate the intertwining space $\mathrm{Hom}_G(\pi, \mathrm{Ind}_{N_S}^G \xi)$. However this intertwining space is infinite-dimensional in general. So we introduce a certain larger group R (See Sect. 2.2 for definition) containing N_S.

A generalized Whittaker model for an admissible representation π of G is a realization of π in the induced module from a closed subgroup which contains N_S. This induced module is a typical example of the reduced generalized Gelfand-Graev representation [**Ya1**]. In this article we investigate the intertwining space

$$GW(\pi;\eta) = \mathrm{Hom}_{(\mathfrak{g},K)}(\pi, C_\eta^\infty(R\backslash G)),$$

where η is an irreducible R-module such that $\eta|_{N_S} \ni \xi$, K is a maximal compact subgroup of G and $\mathfrak{g}$ is the Lie algebra of G. A function in this realization is called a *generalized Whittaker function* for π. The important problem is to determine the dimension of the intertwining space $GW(\pi;\eta)$ and to obtain an explicit formula for generalized Whittaker functions.

We want to discuss this problem for the discrete series representations of G. The group G has three types of discrete series:
(1) Holomorphic and anti-holomorphic discrete series representations.
(2) Two discrete series representations which have ordinary Whittaker

models (cf. [**Kos**]). They are called "Large" discrete series representations (cf. [**Vog**]).
(3) Two non-holomorphic discrete series representations which does not have ordinary Whittaker models. We call them "Middle" discrete series representations.
Naturally we are interested in three types of generalized Whittaker functions for each discrete series representations. Because Yamashita [**Ya2**] determined the embeddings of these discrete series representations into P_J-principal series representations, we firstly consider generalized Whittaker functions for this type of representations. Our first result is a simple integral representation of the generalized Whittaker functions for P_J-principal series representations with corner K-types (Theorems 8.14 and 8.15). We remark that the generalized Whittaker functions for these representations are the same kind of confluent hypergeometric function in case of the large discrete series which is stated below.

Let π_Λ be the discrete series representation with the Harish-Chandra parameter Λ (See Section 9 for definition). Next we consider generalized Whittaker functions for each π_Λ^* with minimal K-type.

First of all, generalized Whittaker functions for the holomorphic discrete series appear in the Fourier expansion of holomorphic Hermitian cusp forms of degree two and their Fourier coefficients are used for construction of Andrianov-Gritsenko-Sugano L-functions. Their generalized Whittaker functions are elementary functions and well known (Theorem 10.2). In this case, Fourier coefficients are indexed by Hermitian matrix H of size two, which parametrize unitary characters of N_S the unipotent radical of the Siegel parabolic subgroup. The Koecher principle [**Koe**] assures that the nonvanishingness of the Fourier coefficient at H implies that H is *positive definite.*

In this paper, we firstly treat the case that characters of N_S are positive definite and the Harish-Chandra parameter belongs to Ξ_{II} (the large discrete series) or belongs to Ξ_{III} (the middle discrete series) respectively. (See Section 9 for definition of Ξ_J, $J = \mathrm{I}, \ldots, \mathrm{VI}$.) We state our first main result.

THEOREM 1.1 (Theorems 10.2, 11.17 and 13.11). *Let π_Λ be the discrete series representation with the Harish-Chandra parameter $\Lambda \in \Xi_{\mathrm{I}} \cup \Xi_{\mathrm{II}} \cup \Xi_{\mathrm{III}}$. We assume that $\eta|_{N_S} \ni \xi = \mathrm{diag}(c_1, c_2)$ is positive definite. Then we have,*

$$\dim_{\mathbb{C}} GW(\pi_\Lambda^*; \eta) = \begin{cases} 1 \ \mathit{or} \ 0 & \cdots \Lambda \in \Xi_{\mathrm{I}}, \\ 2 \ \mathit{or} \ 0 & \cdots \Lambda \in \Xi_{\mathrm{II}}, \\ 0 & \cdots \Lambda \in \Xi_{\mathrm{III}}. \end{cases}$$

The generalized Whittaker functions which appear in the Fourier expansion of automorphic forms are slowly increasing. So, we define $GW(\pi;\eta)^{\mathrm{mod}}$ the subspace of $GW(\pi;\eta)$ which take values in the space of functions with *moderate growth.*

Our second main result is an explicit formula of generalized Whittaker functions and multiplicity one theorem for the large discrete series.

THEOREM 1.2 (Theorem 11.18). 1. *Under the same setting in Theorem 1.1, we assume that* $\dim_{\mathbb{C}} GW(\pi_\Lambda^*;\eta)$ *is not zero. Then we have,*

$$\dim_{\mathbb{C}} GW(\pi_\Lambda^*;\eta)^{\mathrm{mod}} = 1.$$

2. *Moreover if* $\Lambda \in \Xi_{\mathrm{II}}$, *then the slowly increasing generalized Whittaker functions for* π_Λ^* *with minimal* K*-type are uniquely determined and are (up to polynomials) described by the partially confluent hypergeometric functions in* $(a_1, a_2) \in (\mathbb{R}_{>0})^2$:

$$\phi_{\alpha,\beta}(a_1,a_2) = e^{-2\pi(c_1a_1^2+c_2a_2^2)} \int_0^1 W(2\pi c_1 a_1^2 t + 2\pi c_2 a_2^2(1-t)) t^{\alpha}(1-t)^{\beta}\, dt.$$

Here, $W(x)$ *is a confluent hypergeometric function with moderate growth, and* α, β *are non-negative integers.*

For the middle discrete series, there is no generalized Whittaker functions other than zero when characters of N_S are definite, on the contrary to the holomorphic and the large discrete series (Theorem 13.11). This phenomenon is an analogue of the Koecher principle. However the signature condition on H is different, we might call it "an inverse Koecher principle".

Thus we are led to treat also the case that characters of N_S are *indefinite* and the representation belongs to the middle discrete series. Our third main result is an explicit formula of generalized Whittaker functions and multiplicity one theorem for the middle discrete series associated with indefinite characters. In this case, R is isomorphic to the group $SU(1,1) \ltimes N_S$. We write $\eta = \chi \otimes \xi$ with $\chi \in SU(1,1)^\wedge$.

THEOREM 1.3 (Theorem 14.14). 1. *Let* π_Λ *be the discrete series representation with the Harish-Chandra parameter* $\Lambda \in \Xi_{\mathrm{III}}$. *We assume that* $\eta|_{N_S} \ni \xi = \mathrm{diag}(c_1, c_2)$ *is an indefinite character with* $c_1 > 0$, $c_2 < 0$. *If* χ *is an irreducible unitary representation of SU(1,1), then we have*

$$\dim_{\mathbb{C}} GW(\pi_\Lambda^*;\chi,\xi) = \begin{cases} 1 & \cdots\ \chi = D_p^{\pm} \text{ for a finite number of } p, \\ 0 & \cdots\ \text{otherwise.} \end{cases}$$

Here, $D_p^{\pm}$ is the discrete series representation of $SU(1,1)$ with Blattner parameter $\pm p$.

2. *Moreover, if $\chi = D_p^{\pm}$ for the above p, the generalized Whittaker functions for π_Λ^* with minimal K-type are (up to Laurent polynomials) described by the function*

$$b(a_1, a_2) = \begin{cases} e^{2\pi(c_1 a_1^2 - c_2 a_2^2)} & \cdots \chi = D_p^+, \\ e^{-2\pi(c_1 a_1^2 - c_2 a_2^2)} & \cdots \chi = D_p^-. \end{cases}$$

Thus we have

$$\dim_{\mathbb{C}} GW(\pi_\Lambda^*; \chi, \xi)^{\mathrm{mod}} = \begin{cases} 0 & \cdots \chi = D_p^+, \\ 1 & \cdots \chi = D_p^-. \end{cases}$$

Here is some historical comments related to our work. In $Sp(2,\mathbb{R})$ case, which has the same restricted root system of type C_2, this problem has considered by several authors. Niwa [**Ni**] and Miyazaki [**Mi**] obtained explicit integral representation of generalized Whittaker functions. Their results are used for studying non-holomorphic Siegel modular forms and their Andrianov L-functions (cf. [**Ho**]). Among others Miyazaki investigated the large discrete series cases.

We obtain a much simpler integral expression of generalized Whittaker functions belonging to the large discrete series representations than that of Miyazaki. Moreover there is no such representation like the middle discrete series for $Sp(2,\mathbb{R})$. It is a new feature peculiar to $SU(2,2)$.

Acknowledgements. The author would like to express his profound gratitude to Professor Takayuki Oda and Dr. Takahiro Hayata for their valuable suggestions and encouragement, and Dr. Takuya Miyazaki for careful reading of the first version of the manuscript.

CHAPTER 2

Generalized Whittaker functions and representation theory of $SU(2,2)$

2. Definition of the generalized Whittaker functions

2.1. The Siegel parabolic subgroup. Let G denote the special unitary group $SU(2,2)$ and G has two realization G_1, G_2 for example.

$$G_1 = \left\{ g \in SL(4,\mathbb{C}) \,\middle|\, {}^t\bar{g} \begin{pmatrix} 1_2 & \\ & -1_2 \end{pmatrix} g = \begin{pmatrix} 1_2 & \\ & -1_2 \end{pmatrix} \right\}, \tag{1}$$

$$G_2 = \left\{ g \in SL(4,\mathbb{C}) \,\middle|\, {}^t\bar{g} \begin{pmatrix} & 1_2 \\ -1_2 & \end{pmatrix} g = \begin{pmatrix} & 1_2 \\ -1_2 & \end{pmatrix} \right\}. \tag{2}$$

Here we denote by 1_2 the unit matrix of size two. We fix the isomorphism $\varphi_{12} : G_1 \ni g \mapsto CgC^{-1} \in G_2$ hereafter. Here,

$$C = \frac{1}{\sqrt{2}} \begin{pmatrix} 1_2 & 1_2 \\ -\sqrt{-1}\cdot 1_2 & \sqrt{-1}\cdot 1_2 \end{pmatrix}$$

and also put $\varphi_{21} = \varphi_{12}^{-1}$.

We use realization G_2 as $G = SU(2,2)$ in this section and take the Siegel parabolic subgroup P_S of G, i.e. maximal parabolic subgroup with abelian unipotent radical.

$$G_2 \supset P_S = \left\{ \begin{pmatrix} * & * \\ 0_2 & * \end{pmatrix} \in G_2 \right\}.$$

We write $L_S \ltimes N_S$ the Levi decomposition of P_S and these subgroups are realized as follows.

$$L_S = \left\{ l(g) = \begin{pmatrix} g & \\ & {}^t\bar{g}^{-1} \end{pmatrix} \,\middle|\, g \in GL(2,\mathbb{C}),\ \det g \in \mathbb{R} \right\},$$

$$N_S = \left\{ n(B) = \begin{pmatrix} 1_2 & B \\ & 1_2 \end{pmatrix} \,\middle|\, {}^t\bar{B} = B \in M_2(\mathbb{C}) \right\} \simeq \mathbb{R}^4.$$

2.2. Characters of N_S. We define a unitary character ξ of N_S associated with a Hermitian matrix $H_\xi = \begin{pmatrix} c_1 & \gamma \\ \bar{\gamma} & c_2 \end{pmatrix}$. For $n = n(B) \in$

N_S, $B = \begin{pmatrix} a & \beta \\ \bar{\beta} & d \end{pmatrix}$,

$$\begin{aligned}\xi(n(B)) = \xi(\begin{pmatrix} 1_2 & B \\ & 1_2 \end{pmatrix}) &= \exp 2\pi\sqrt{-1}\mathrm{tr}\{\begin{pmatrix} c_1 & \gamma \\ \bar{\gamma} & c_2 \end{pmatrix}\begin{pmatrix} a & \beta \\ \bar{\beta} & d \end{pmatrix}\} \\ &= \exp 2\pi\sqrt{-1}(c_1 a + c_2 d + \gamma\bar{\beta} + \bar{\gamma}\beta) \in \mathbb{C}^{(1)}.\end{aligned}$$

Hereafter we assume that ξ is nondegenerate, i.e. $\det H_\xi \neq 0$. Consider the action of L_S on N_S by conjugation. Then we take the identity component of the stabilizer subgroup of the character ξ under the induced action of L_S on the character group $\widehat{N_S}$ and denote it as $SU(\xi)$,

$$SU(\xi) = \mathrm{Stab}_{L_S}(\xi)^\circ \subset L_S.$$

Then

$$SU(\xi) = \left\{ \begin{pmatrix} g & \\ & {}^t\bar{g}^{-1} \end{pmatrix} \,\middle|\, \begin{array}{l} g \in GL(2,\mathbb{C}),\ \det g \in \mathbb{R} \\ {}^t\bar{g}H_\xi g = H_\xi \end{array} \right\}^\circ.$$

Hence

$$SU(\xi) \simeq \begin{cases} SU(2) & \text{for definite } H_\xi, \\ SU(1,1) & \text{for indefinite } H_\xi. \end{cases}$$

We mainly treat the case when H_ξ is definite in this article, i.e. $SU(\xi) \simeq SU(2)$.

Define the subgroup R_ξ of L_S,

$$R_\xi = SU(\xi) \ltimes N_S$$

and take an irreducible unitary representation χ of $SU(\xi)$. For $\xi \in \widehat{N_S}$ and $\chi \in SU(\xi)^\wedge$ as above, we can define a well-defined R_ξ-module $\eta = \chi \otimes \xi$.

2.3. Generalized Whittaker models and functions.

DEFINITION 2.1. For an admissible representation (π, H_π), define $GW(\pi;\eta)$ the space of the following intertwining operators,

$$GW(\pi;\chi,\xi) = GW(\pi;\eta) = \mathrm{Hom}_{(\mathfrak{g},K)}(H_\pi, C_\eta^\infty(R_\xi\backslash G)).$$

Here,

$$C_\eta^\infty(R_\xi\backslash G) = \left\{ f\colon G \overset{C^\infty}{\to} V_\chi \mid f(rg) = \eta(r)f(g) \quad \forall (r,g) \in R \times G \right\}.$$

We also define $C_\eta^\infty(R_\xi\backslash G)^{\mathrm{mod}}$ the space of functions with *moderate growth* in $C_\eta^\infty(R_\xi\backslash G)$ (See Wallach [**Wal**, Sect. 8]).

DEFINITION 2.2. Take $T\,(\neq 0) \in GW(\pi;\eta)$ and multiplicity one K-type (τ^*, W_{τ^*}) of π. Here, τ^* is the contragredient representation of τ. Then we fix a K-equivariant injection $\iota\colon W_{\tau^*} \to H_\pi$ and consider the image of $\iota(v^*)$, $v^* \in W_{\tau^*}$,

$$T(\iota_{\tau^*}(v^*))(g) = \langle v^*, \Phi_{\pi,\tau}(g)\rangle.$$

Here, $\langle , \rangle$ is the canonical pairing on $W_\tau \times W_{\tau^*}$. This equality defines (up to a constant) a function $\Phi_{\pi,\tau}(g) \in C^\infty_{\eta,\tau}(R\backslash G/K)$. Here,

$$C^\infty_\eta(R\backslash G/K)) = \Big\{ f\colon G \overset{C^\infty}{\to} V_\chi \otimes W_\tau \mid f(rgk) = \eta(r) \otimes \tau(k^{-1}) f(g) \\ \forall (r,g,k) \in R \times G \times K \Big\}.$$

We call $\Phi_{\pi,\tau}$ a generalized Whittaker function (= GWF for short) for π with K-type τ^*.

3. Structure theory for $SU(2,2)$ and its Lie algebra

3.1. Lie groups and their Lie algebras. Let G denote the special unitary group $SU(2,2)$ realized as

$$G = \{g \in SL(4,\mathbb{C}) | {}^t\bar{g} I_{2,2} g = I_{2,2}\}$$

where $I_{2,2} = \mathrm{diag}(1,1,-1,-1)$ and we denote by ${}^tg$ and $\bar{g}$ the transpose and the complex conjugation of g respectively, i.e. realization G_1. We fix a maximal compact subgroup K of G as follows:

$$K = \left\{ \begin{pmatrix} u_1 & \\ & u_2 \end{pmatrix} \middle| \, u_1, u_2 \in U(2),\ \det(u_1)\det(u_2) = 1 \right\}.$$

The Lie algebra $\mathfrak{g}$ of G is expressed as follows:

$$\mathfrak{g} = \left\{ \begin{pmatrix} X_1 & X_3 \\ {}^t\bar{X}_3 & X_2 \end{pmatrix} \middle| \begin{array}{l} {}^t\bar{X}_i = -X_i\ (i=1,2),\quad X_3 \in M_2(\mathbb{C}), \\ \mathrm{tr}(X_1 + X_2) = 0 \end{array} \right\}.$$

The Lie algebra of K is:

$$\mathfrak{k} = \left\{ \begin{pmatrix} X_1 & \\ & X_2 \end{pmatrix} \in \mathfrak{g} \right\}, \qquad \mathfrak{p} = \left\{ \begin{pmatrix} & X_3 \\ {}^t\bar{X}_3 & \end{pmatrix} \in \mathfrak{g} \right\}.$$

Then we have the Cartan decomposition: $\mathfrak{g} = \mathfrak{k} + \mathfrak{p}$. We fix a maximal abelian subalgebra $\mathfrak{a}$ of $\mathfrak{p}$ as follows:

$$\mathfrak{a} = \mathbb{R}H_1 + \mathbb{R}H_2 \quad \text{with } H_1 = \begin{pmatrix} & & 1 & \\ & & & 0 \\ 1 & & & \\ & 0 & & \end{pmatrix}, \ H_2 = \begin{pmatrix} & & 0 & \\ & & & 1 \\ 0 & & & \\ & 1 & & \end{pmatrix}.$$

The Lie group $A = \exp(\mathfrak{a})$ is the identity component of a maximal $\mathbb{R}$-split torus in G.

We identify A with $(\mathbb{R}_{>0})^2$ by,

$$A \ni (a_1, a_2) = \exp(\log(a_1)H_1 + \log(a_2)H_2) \quad \text{with} \quad (a_1, a_2) \in (0, \infty)^{\oplus 2}.$$

Let Σ denote the restricted root system of $(\mathfrak{g}, \mathfrak{a})$. Then Σ is of type C_2 and we choose positive system Σ^+ as

$$\Sigma^+ = \{2\alpha_1,\, 2\alpha_2,\, \alpha_1 + \alpha_2,\, \alpha_1 - \alpha_2\} \text{ with } \alpha_i(H_j) = \delta_{i,j}\ (i, j = 1, 2).$$

We set

$$E_1 = \sqrt{-1}\begin{pmatrix} 1 & & -1 & \\ & 0 & & 0 \\ 1 & & -1 & \\ & 0 & & 0 \end{pmatrix}, \quad E_2 = \sqrt{-1}\begin{pmatrix} 0 & & 0 & \\ & 1 & & -1 \\ 0 & & 0 & \\ & 1 & & -1 \end{pmatrix},$$

$$E_3 = \frac{1}{2}\begin{pmatrix} & 1 & & -1 \\ -1 & & 1 & \\ & 1 & & -1 \\ -1 & & 1 & \end{pmatrix}, \quad E_4 = \frac{\sqrt{-1}}{2}\begin{pmatrix} & 1 & & -1 \\ 1 & & -1 & \\ & 1 & & -1 \\ 1 & & -1 & \end{pmatrix},$$

$$E_5 = \frac{1}{2}\begin{pmatrix} & 1 & & 1 \\ -1 & & 1 & \\ & 1 & & 1 \\ 1 & & -1 & \end{pmatrix}, \quad E_6 = \frac{\sqrt{-1}}{2}\begin{pmatrix} & 1 & & 1 \\ 1 & & -1 & \\ & 1 & & 1 \\ -1 & & 1 & \end{pmatrix}.$$

Then each root space $\mathfrak{g}_\alpha$, $\alpha \in \Sigma^+$ is given as follows:

$$\mathfrak{g}_{2\alpha_1} = \mathbb{R}E_1, \quad \mathfrak{g}_{2\alpha_2} = \mathbb{R}E_2, \quad \mathfrak{g}_{\alpha_1+\alpha_2} = \mathbb{R}E_3 + \mathbb{R}E_4,$$

and

$$\mathfrak{g}_{\alpha_1-\alpha_2} = \mathbb{R}E_5 + \mathbb{R}E_6.$$

Put

$$\mathfrak{n}_S = \mathfrak{g}_{2\alpha_1} + \mathfrak{g}_{2\alpha_2} + \mathfrak{g}_{\alpha_1+\alpha_2}, \tag{3}$$

$$\mathfrak{n} = \mathfrak{n}_S + \mathfrak{g}_{\alpha_1-\alpha_2}. \tag{4}$$

Then we have the Iwasawa decomposition $\mathfrak{g} = \mathfrak{n} + \mathfrak{a} + \mathfrak{k}$ and $\mathfrak{n}_S = \mathrm{Lie}(N_S)$, N_S is the nilpotent radical of the Siegel parabolic subgroup P_S in **realization** G_1. We only use realization G_1 hereafter.

Set $E_{-i} = \theta E_i = -{}^t\bar{E}_i$, $(i = 1, \ldots, 6)$. Then $\mathfrak{g}_{-2\alpha_1} = \mathbb{R}E_{-1}$ etc.

We also fix a basis of $\mathfrak{k}_{\mathbb{C}}$ as

$$\mathfrak{k}_{\mathbb{C}} = \mathbb{C}h^1 + \mathbb{C}h^2 + \mathbb{C}e^1_+ + \mathbb{C}e^1_- + \mathbb{C}e^2_+ + \mathbb{C}e^2_- + \mathbb{C}I_{2,2}, \tag{5}$$

here,

$$h^1 = \begin{pmatrix} h & \\ & 0_2 \end{pmatrix}, h^2 = \begin{pmatrix} 0_2 & \\ & h \end{pmatrix}, e^1_\pm = \begin{pmatrix} e_\pm & \\ & 0_2 \end{pmatrix}, e^2_\pm = \begin{pmatrix} 0_2 & \\ & e_\pm \end{pmatrix}, \tag{6}$$

and

$$h = \begin{pmatrix} 1 & \\ & -1 \end{pmatrix}, \quad e_+ = \begin{pmatrix} 0 & 1 \\ & 0 \end{pmatrix}, \quad e_- = \begin{pmatrix} & 0 \\ 1 & \end{pmatrix}. \tag{7}$$

The set $\{h, e_+, e_-\}$ is called a $\mathfrak{sl}_2$-triple. Take a compact Cartan subalgebra $\mathfrak{t}$ defined by

$$\mathfrak{t} = \mathbb{R}\sqrt{-1}h^1 + \mathbb{R}\sqrt{-1}h^2 + \mathbb{R}\sqrt{-1}I_{2,2}. \tag{8}$$

The root system Δ of $(\mathfrak{g}_{\mathbb{C}}, \mathfrak{t}_{\mathbb{C}})$, of type A_3, is expressed by,

$$\Delta = \{[\pm 2, 0; 0], [0, \pm 2; 0], [\pm 1, \pm 1; \pm 2]\}.$$

where $\beta = [r, s; u]$ means that $r = \beta(h^1)$, $s = \beta(h^2)$ and $u = \beta(I_{2,2})$. Four roots $\pm[2,0;0]$, $\pm[0,2;0]$ form the subsystem $\Delta_c \subset \Delta$ of compact roots. We fix a positive system:

$$\Delta_c^+ = \{[2,0;0], [0,2;0]\} \subset \Delta_c. \tag{9}$$

3.2. The stabilizer of ξ. We describe 3-dimensional Lie subalgebra $\mathfrak{su}(\xi) = \mathrm{Lie}(SU(\xi))$ of $\mathfrak{g}$. If S is a subgroup of G, we denote it by S' in realization G_2, (see (2)). Define the map σ by

$$\sigma : SU(\xi)' \ni \begin{pmatrix} g & \\ & {}^t\bar{g}^{-1} \end{pmatrix} \mapsto g \in GL(2, \mathbb{C}).$$

Then

$$\sigma(SU(\xi)') = \{g \in SL(2,\mathbb{C}) | \, {}^t\bar{g} H_\xi g = H_\xi\},$$

and

$$\mathrm{Lie}(\sigma(SU(\xi)')) = \{X \in \mathfrak{sl}(2,\mathbb{C}) | \, {}^t\bar{X} H_\xi + H_\xi X = 0\}.$$

LEMMA 3.1. *For* $\xi = \begin{pmatrix} c_1 & \gamma \\ \bar{\gamma} & c_2 \end{pmatrix}$ *such that* $c_2 \neq 0$ *and* $\det H_\xi \neq 0$, *Put*

$$Z_1 = \sqrt{-1}\begin{pmatrix} 1 & 0 \\ -2c_2^{-1}\bar{\gamma} & -1 \end{pmatrix}, \quad Z_2 = \sqrt{-1}\begin{pmatrix} \bar{\gamma} & c_2 \\ c_2^{-1}(\det H_\xi - \bar{\gamma}^2) & -\bar{\gamma} \end{pmatrix},$$
$$Z_3 = \begin{pmatrix} -\bar{\gamma} & -c_2 \\ c_2^{-1}(\det H_\xi + \bar{\gamma}^2) & \bar{\gamma} \end{pmatrix}. \tag{10}$$

Then we can take $\{Z_1, Z_2, Z_3\}$ *as a basis of* $\mathrm{Lie}(\sigma(SU(\xi)'))$ *over* $\mathbb{R}$, *i.e.*

$$[Z_1, Z_2] = 2Z_3, \quad [Z_2, Z_3] = 2(\det H_\zeta)Z_1, \quad [Z_3, Z_1] = 2Z_2.$$

PROOF. We can easily check it by direct computation. ☐

Next we set $X_i = d\varphi_{21} \circ d\sigma^{-1}(Z_i)$ $(i = 1, 2, 3)$.

LEMMA 3.2.

$$\begin{aligned} X_1 =& I_0 - 2c_2^{-1}\mathrm{Re}(\gamma)E_{-6} + 2c_2^{-1}\mathrm{Im}(\gamma)E_{-5}, \\ X_2 =& \mathrm{Re}(\gamma)I_0 + c_2^{-1}(\det H_\xi - \mathrm{Re}(\gamma^2))E_{-6} + c_2E_6 + c_2^{-1}\mathrm{Im}(\gamma^2)E_{-5} \\ &+ \mathrm{Im}(\gamma)(H_1 - H_2), \\ X_3 =& \mathrm{Im}(\gamma)I_0 - c_2^{-1}(\det H_\xi + \mathrm{Re}(\gamma^2))E_{-5} - c_2E_5 - c_2^{-1}\mathrm{Im}(\gamma^2)E_{-6} \\ &- \mathrm{Re}(\gamma)(H_1 - H_2). \end{aligned}$$

Here, $I_0 = \sqrt{-1}\mathrm{diag}(1,-1,1,-1) = \sqrt{-1}(h^1 + h^2) \in \mathfrak{k}$.

PROOF. Note that

$$d\varphi_{21} \circ d\sigma^{-1}(X) = \frac{1}{2}\begin{pmatrix} X - {}^t\bar{X} & X + {}^t\bar{X} \\ X + {}^t\bar{X} & X - {}^t\bar{X} \end{pmatrix}.$$

We readily obtain the assertion. □

Thus we can take $\{X_1, X_2, X_3\}$ as a basis of $\mathfrak{su}(\xi)$ over $\mathbb{R}$ and

$$[X_1, X_2] = 2X_3, \quad [X_2, X_3] = 2(\det H_\xi)X_1, \quad [X_3, X_1] = 2X_2.$$

We put

$$X_5 = X_3 - \mathrm{Im}(\gamma)X_1, \quad X_6 = X_2 - \mathrm{Re}(\gamma)X_1, \tag{11}$$

and

$$Y_i = E_i + E_{-i} \in \mathfrak{k}_{\mathbb{C}} \ (i = 5, 6), \quad \triangle = (c_1a_1^2 - c_2a_2^2).$$

We also note that

$$Y_5 + \sqrt{-1}Y_6 = -2(e_-^1 + e_-^2), \quad Y_5 - \sqrt{-1}Y_6 = 2(e_+^1 + e_+^2).$$

This implies,

LEMMA 3.3. *For the basis of* $\mathfrak{g}_{\alpha_1-\alpha_2}$*, we have*

$$E_5 = \frac{a_1a_2}{\triangle}\left\{\mathrm{Ad}(a^{-1})X_5 + c_1\left(\frac{a_1}{a_2}\right)Y_5 + \mathrm{Re}(\gamma)(H_1 - H_2)\right\}, \tag{12}$$

$$E_6 = -\frac{a_1a_2}{\triangle}\left\{\mathrm{Ad}(a^{-1})X_6 - c_1\left(\frac{a_1}{a_2}\right)Y_6 - \mathrm{Im}(\gamma)(H_1 - H_2)\right\}. \tag{13}$$

Then we have the Iwasawa-Cartan type decomposition of $\mathfrak{g}$.

LEMMA 3.4.

$$\mathfrak{g} = \mathrm{Ad}(a^{-1})(\mathfrak{su}(\xi) + \mathfrak{n}_S) + \mathfrak{a} + \mathfrak{k}.$$

3.3. Iwasawa decomposition of elements in $\mathfrak{p}_{\pm}$. Let $X_{ij} = (\delta_{ip}\delta_{jq})_{pq}$ and $H_{ij} = \sqrt{-1}(X_{ii} - X_{jj})$. Put

$$\mathfrak{p}_{+} = \mathbb{C}X_{13} + \mathbb{C}X_{23} + \mathbb{C}X_{14} + \mathbb{C}X_{24},$$

$$\mathfrak{p}_{-} = \mathbb{C}X_{31} + \mathbb{C}X_{41} + \mathbb{C}X_{32} + \mathbb{C}X_{42}.$$

Then $\mathfrak{g}_{\mathbb{C}} = \mathfrak{k}_{\mathbb{C}} + \mathfrak{p}_{\mathbb{C}}$ and $\mathfrak{p}_{\mathbb{C}} = \mathfrak{p}_{+} + \mathfrak{p}_{-}$. We have the following formula by direct computation.

LEMMA 3.5.

$$\begin{aligned}
2X_{13} &= \sqrt{-1}E_1 + H_1 - \sqrt{-1}H_{13},\\
2X_{23} &= E_3 + E_5 + \sqrt{-1}(E_4 + E_6) + 2e^1_{-},\\
2X_{14} &= -E_3 + E_5 + \sqrt{-1}(E_4 - E_6) - 2e^2_{+},\\
2X_{24} &= \sqrt{-1}E_2 + H_2 - \sqrt{-1}H_{24},\\
2X_{31} &= -\sqrt{-1}E_1 + H_1 + \sqrt{-1}H_{13},\\
2X_{41} &= -E_3 + E_5 - \sqrt{-1}(E_4 - E_6) + 2e^2_{-},\\
2X_{32} &= E_3 + E_5 - \sqrt{-1}(E_4 + E_6) - 2e^1_{+},\\
2X_{42} &= -\sqrt{-1}E_2 + H_2 + \sqrt{-1}H_{24}.
\end{aligned}$$

4. Representations of K

4.1. Irreducible K-modules. We recall the parametrization of the irreducible representations of $K \simeq S(U(2) \times U(2))$. Since $\mathfrak{k}_{\mathbb{C}} \simeq \mathfrak{sl}(2,\mathbb{C}) \oplus \mathfrak{sl}(2,\mathbb{C}) \oplus \mathbb{C}$, we begin with irreducible finite-dimensional representations of $\mathfrak{sl}(2,\mathbb{C})$. For every integer $d \geq 0$, $\mathfrak{sl}(2,\mathbb{C})$ has a unique (up to equivalence) irreducible complex linear representation τ_d on a space V_d of dimension $d+1$. Then V_d has a basis $\{f_j = f_j^{(d)} | 0 \leq j \leq d\}$ consisting of weight vectors satisfying

$$\begin{aligned}
\tau_d(h)f_j &= (2j - d)f_j,\\
\tau_d(e_{+})f_j &= (d - j)f_{j+1},\\
\tau_d(e_{-})f_j &= jf_{j-1}.
\end{aligned}$$

Here, $\{h, e_{+}, e_{-}\}$ is a $\mathfrak{sl}_2$-triple. This representation (τ_d, V_d) determines the representation of $\mathfrak{su}(2)$ by restriction and that of $SU(2)$ by exponentiation, isomorphic to the symmetric tensor representation $\mathrm{Sym}^d(\mathbb{C}^2)$.

We will give the parametrizations of irreducible K-modules. Let μ be a Δ_c^{+}-dominant, integral linear form on $\mathfrak{t}$, i.e., $\mu = [\mu_1, \mu_2; \mu_3]$ with integers μ_j such that $\mu_1, \mu_2 \geq 0$ and $\mu_1 + \mu_2 + \mu_3 \in 2\mathbb{Z}$. Let $\mathfrak{z} = \mathbb{C}I_{2,2}$ be the center of $\mathfrak{t}_{\mathbb{C}}$ and $\mathfrak{t}'_{\mathbb{C}} = [\mathfrak{t}_{\mathbb{C}}, \mathfrak{t}_{\mathbb{C}}] \simeq \mathfrak{sl}(2,\mathbb{C}) \oplus \mathfrak{sl}(2,\mathbb{C})$ the derived algebra. Then we have $\mathfrak{t}_{\mathbb{C}} = \mathfrak{z} \oplus \mathfrak{t}'_{\mathbb{C}}$.

The irreducible $\mathfrak{k}_{\mathbb{C}}$-module (τ_μ, W_μ) with highest weight $\mu = [r, s; u]$ is realized on $W_\mu = V_r \otimes V_s$ with the action

$$\tau(\begin{pmatrix} Y_1 & \\ & Y_2 \end{pmatrix}) = \mathrm{Sym}^r(Y_1) \otimes \mathrm{id}_{V_s} + \mathrm{id}_{V_r} \otimes \mathrm{Sym}^s(Y_2)$$
$$(Y_1, Y_2 \in \mathfrak{sl}(2, \mathbb{C})), \tag{14}$$
$$\tau(I_{2,2}) = u \cdot \mathrm{id}_{W_\mu}. \tag{15}$$

Let us write $W_{rs} = W_\mu$ and choose the standard basis of W_{rs} by using those of V_r and V_s, as

$$\{f_{kl} = f_{kl}^{(rs)} = f_k^{(r)} \otimes f_l^{(s)} | 0 \le k \le r,\ 0 \le l \le s\} \tag{16}$$

We summarize this subsection.

LEMMA 4.1. *We have the isomorphism*

$$\widehat{K} \simeq \{[r, s; u] \mid r, s \in \mathbb{Z}_{\ge 0},\ u \in \mathbb{Z},\ r + s + u \in 2\mathbb{Z}\}.$$

Let $\tau = [r, s; u] \in \widehat{K}$. *Then,*

$$\tau(h^1) f_{kl} = (2k - r) f_{kl}, \qquad \tau(h^2) f_{kl} = (2l - s) f_{kl},$$
$$\tau(e_+^1) f_{kl} = (r - k) f_{k+1,l}, \qquad \tau(e_+^2) f_{kl} = (s - l) f_{k,l+1},$$
$$\tau(e_-^1) f_{kl} = k f_{k-1,l}, \qquad \tau(e_-^2) f_{kl} = l f_{k,l-1},$$
$$\tau(I_{2,2}) f_{kl} = u f_{kl}.$$

4.2. Adjoint representation of K. We consider the adjoint representation of K restricted to $\mathfrak{p}_{\mathbb{C}}$ and write it by Ad. Set $\mathrm{Ad}^\pm = \mathrm{Ad}|_{\mathfrak{p}_\pm}$.

PROPOSITION 4.2 ([**Ya2**, Sect. 4.2]). *The representations* $\mathrm{Ad}^\pm$ *are irreducible. In fact,*

$$\mathrm{Ad}^+ \simeq [1, 1; 2], \quad \mathrm{Ad}^- \simeq [1, 1; -2].$$

as K-modules. The K-isomorphism are given by

$$\iota_+ : (X_{23}, X_{13}, X_{24}, X_{14}) \mapsto (f_{00}^{(11)}, f_{10}^{(11)}, -f_{01}^{(11)}, -f_{11}^{(11)}), \tag{17}$$
$$\iota_- : (X_{41}, X_{31}, X_{42}, X_{32}) \mapsto (f_{00}^{(11)}, f_{01}^{(11)}, -f_{10}^{(11)}, -f_{11}^{(11)}). \tag{18}$$

Let $P_r^\pm$ be the projectors from $V_r \otimes V_1$ to $V_{r\pm1}$ defined by

$$P_r^+(f_r^{(r)} \otimes f_1^{(1)}) = f_{r+1}^{(r+1)}, \quad P_r^+(f_{r-1}^{(r)} \otimes f_1^{(1)} - f_r^{(r)} \otimes f_0^{(1)}) = 0,$$
$$P_r^-(f_r^{(r)} \otimes f_1^{(1)}) = 0, \quad P_r^+(f_{r-1}^{(r)} \otimes f_1^{(1)} - f_r^{(r)} \otimes f_0^{(1)}) = (r+1) f_{r-1}^{(r-1)}.$$

Then we have,

LEMMA 4.3. *For* $0 \le j \le r$, *we have*

$$P_r^+(f_j^{(r)} \otimes f_0^{(1)}) = f_j^{(r+1)}, \qquad P_r^+(f_j^{(r)} \otimes f_1^{(1)}) = f_{j+1}^{(r+1)},$$
$$P_r^-(f_j^{(r)} \otimes f_0^{(1)}) = -j f_{j-1}^{(r-1)}, \quad P_r^-(f_j^{(r)} \otimes f_1^{(1)}) = (r - j) f_j^{(r-1)}.$$

Let us consider the tensor product $W_\tau \otimes \mathfrak{p}_\pm$. For $\tau = [r,s;u]$, $W_\tau = W_{rs} = V_r \otimes V_s$, we see that $W_{rs} \otimes \mathfrak{p}_\pm$ decompose into four $[\mathfrak{k},\mathfrak{k}]$-submodules $W_{r\pm1,s\pm1}$. We define the projectors

$$P_{rs}^{(\epsilon_1,\epsilon_2)} : W_{r,s} \otimes \mathfrak{p}_+ \to W_{r+\epsilon_1,s+\epsilon_2}, \tag{19}$$

$$\bar{P}_{rs}^{(\epsilon_1,\epsilon_2)} : W_{r,s} \otimes \mathfrak{p}_- \to W_{r+\epsilon_1,s+\epsilon_2}. \tag{20}$$

LEMMA 4.4. *Put* $P^{(\epsilon_1,\epsilon_2)} = P_{rs}^{(\epsilon_1,\epsilon_2)}$ *and* $\bar{P}^{(\epsilon_1,\epsilon_2)} = \bar{P}_{rs}^{(\epsilon_1,\epsilon_2)}$. *Then we have,*

$$\begin{aligned}
P^{(+,+)}(f_{kl}\otimes X_{13}) &= -\bar{P}^{(+,+)}(f_{kl}\otimes X_{42}) = f_{k+1,l},\\
P^{(+,+)}(f_{kl}\otimes X_{23}) &= \bar{P}^{(+,+)}(f_{kl}\otimes X_{41}) = f_{k,l},\\
P^{(+,+)}(f_{kl}\otimes X_{14}) &= \bar{P}^{(+,+)}(f_{kl}\otimes X_{32}) = -f_{k+1,l+1},\\
P^{(+,+)}(f_{kl}\otimes X_{24}) &= -\bar{P}^{(+,+)}(f_{kl}\otimes X_{31}) = -f_{k,l+1},\\
P^{(-,-)}(f_{kl}\otimes X_{13}) &= -\bar{P}^{(-,-)}(f_{kl}\otimes X_{42}) = (k-r)lf_{k,l-1},\\
P^{(-,-)}(f_{kl}\otimes X_{23}) &= \bar{P}^{(-,-)}(f_{kl}\otimes X_{41}) = klf_{k-1,l-1},\\
P^{(-,-)}(f_{kl}\otimes X_{14}) &= \bar{P}^{(-,-)}(f_{kl}\otimes X_{32}) = (k-r)(s-l)f_{kl},\\
P^{(-,-)}(f_{kl}\otimes X_{24}) &= -\bar{P}^{(-,-)}(f_{kl}\otimes X_{31}) = k(s-l)f_{k-1,l},\\
P^{(+,-)}(f_{kl}\otimes X_{13}) &= -\bar{P}^{(+,-)}(f_{kl}\otimes X_{42}) = (-l)f_{k+1,l-1},\\
P^{(+,-)}(f_{kl}\otimes X_{23}) &= \bar{P}^{(+,-)}(f_{kl}\otimes X_{41}) = (-l)f_{k,l-1},\\
P^{(+,-)}(f_{kl}\otimes X_{14}) &= \bar{P}^{(+,-)}(f_{kl}\otimes X_{32}) = (l-s)f_{k+1,l},\\
P^{(+,-)}(f_{kl}\otimes X_{24}) &= -\bar{P}^{(+,-)}(f_{kl}\otimes X_{31}) = (l-s)f_{k,l},\\
P^{(-,+)}(f_{kl}\otimes X_{13}) &= -\bar{P}^{(-,+)}(f_{kl}\otimes X_{42}) = (r-k)f_{k,l},\\
P^{(-,+)}(f_{kl}\otimes X_{23}) &= \bar{P}^{(-,+)}(f_{kl}\otimes X_{41}) = (-k)f_{k-1,l},\\
P^{(-,+)}(f_{kl}\otimes X_{14}) &= \bar{P}^{(-,+)}(f_{kl}\otimes X_{32}) = (k-r)f_{k,l+1},\\
P^{(-,+)}(f_{kl}\otimes X_{24}) &= -\bar{P}^{(-,+)}(f_{kl}\otimes X_{31}) = kf_{k-1,l+1}.
\end{aligned}$$

PROOF. These are consequence of Proposition 4.2 and Lemma 4.3. □

5. Irreducible $SU(\xi)$-modules

We only treat the case of $\gamma = 0$ for simplicity. By using Lemma 3.2 we have:

$$X_5 \pm \sqrt{-1}X_6 = X_3 \pm \sqrt{-1}X_2.$$

and

$$\begin{aligned}
[-\sqrt{-1}X_1,\, X_3 + \sqrt{-1}X_2] &= 2(X_3 + \sqrt{-1}X_2),\\
[-\sqrt{-1}X_1,\, X_3 - \sqrt{-1}X_2] &= -2(X_3 - \sqrt{-1}X_2),\\
[X_3 + \sqrt{-1}X_2,\, X_3 - \sqrt{-1}X_2] &= -4c_1c_2(-\sqrt{-1}X_1).
\end{aligned}$$

Therefore, we define the following basis of $\mathfrak{su}(\xi)_{\mathbb{C}}$:
(1) If $c_1 > 0$ and $c_2 > 0$, i.e. ξ is definite, then

$$\{-\sqrt{-1}X_1,\ -(4c_1c_2)^{-\frac{1}{2}}(X_3+\sqrt{-1}X_2),\quad (4c_1c_2)^{-\frac{1}{2}}(X_3-\sqrt{-1}X_2)\},$$

(2) If $c_1 < 0$ and $c_2 < 0$, i.e. ξ is definite, then

$$\{-\sqrt{-1}X_1,\quad (4c_1c_2)^{-\frac{1}{2}}(X_3+\sqrt{-1}X_2),\ -(4c_1c_2)^{-\frac{1}{2}}(X_3-\sqrt{-1}X_2)\},$$

(3) If $c_1 > 0$ and $c_2 < 0$, i.e. ξ is indefinite, then

$$\{-\sqrt{-1}X_1,\quad (-4c_1c_2)^{-\frac{1}{2}}(X_3+\sqrt{-1}X_2),\quad (-4c_1c_2)^{-\frac{1}{2}}(X_3-\sqrt{-1}X_2)\},$$

(4) If $c_1 < 0$ and $c_2 > 0$, i.e. ξ is indefinite, then

$$\{-\sqrt{-1}X_1,\ -(-4c_1c_2)^{-\frac{1}{2}}(X_3+\sqrt{-1}X_2),\ -(-4c_1c_2)^{-\frac{1}{2}}(X_3-\sqrt{-1}X_2)\}.$$

We see that each of them forms a $\mathfrak{sl}_2$-triple.

5.1. The case of $SU(\xi) \simeq SU(2)$. We treat the case of ξ is positive definite.

LEMMA 5.1. *If ξ is positive definite, take an irreducible $SU(\xi)$-module (i.e. $SU(2)$-module) $\chi = \mathrm{Sym}^d(\mathbb{C}^2)$ with representation space V_χ. Then V_χ has a standard basis $\{v_j \mid j = 0, \dots, d\}$ which satisfies the following:*

$$\begin{aligned}\chi(-\sqrt{-1}X_1)v_j &= (2j-d)v_j,\\ \chi(X_3+\sqrt{-1}X_2)v_j &= -2\sqrt{c_1c_2}\,(d-j)v_{j+1},\\ \chi(X_3-\sqrt{-1}X_2)v_j &= 2\sqrt{c_1c_2}\,jv_{j-1}.\end{aligned}$$

Therefore we put

$$X_\pm = (2\sqrt{c_1c_2})^{-1}(X_3 \pm \sqrt{-1}X_2).$$

5.2. The case of $SU(\xi) \simeq SU(1,1)$. We treat the case of $c_1 > 0$ and $c_2 < 0$. For $\epsilon \in \{0,1\}$ and $\mu \in \mathbb{C}$, we denote by $(\chi_{\epsilon,\mu}, H_{\epsilon,\mu})$ the non-unitary principal series representation of $SU(1,1)$. Define

$$\mathbb{Z}_\epsilon = \{j \in \mathbb{Z} \mid j \equiv \epsilon \pmod 2\}.$$

PROPOSITION 5.2. *There is a standard basis* $\{u_j \mid j \in \mathbb{Z}_\epsilon\}$ *of* $(\chi_{\epsilon,\mu}, H_{\epsilon,\mu})$ *with action:*

$$\chi_{\epsilon,\mu}(h)u_j = ju_j,$$
$$\chi_{\epsilon,\mu}(e_+)u_j = \frac{1}{2}(j+1+\mu)u_{j+2},$$
$$\chi_{\epsilon,\mu}(e_-)u_j = \frac{1}{2}(-j+1+\mu)u_{j-2},$$

in other words,

$$\chi_{\epsilon,\mu}(-\sqrt{-1}X_1)u_j = ju_j,$$
$$\chi_{\epsilon,\mu}(X_3+\sqrt{-1}X_2)u_j = \sqrt{-c_1c_2}(j+1+\mu)u_{j+2},$$
$$\chi_{\epsilon,\mu}(X_3-\sqrt{-1}X_2)u_j = \sqrt{-c_1c_2}(-j+1+\mu)u_{j-2}.$$

Let ϵ' be 0 or 1 different from ϵ. For non-negative integer $m \in \mathbb{Z}_{\epsilon'}$, we define

$$V_{m+1} = \bigoplus_{\substack{j\in\mathbb{Z}_\epsilon \\ j\geq m+1}} \mathbb{C}u_j \qquad : \text{lowest weight module,}$$

$$\bar{V}_{m+1} = \bigoplus_{\substack{j\in\mathbb{Z}_\epsilon \\ j\leq -m-1}} \mathbb{C}u_j \qquad : \text{highest weight module.}$$

PROPOSITION 5.3 (Unitary dual of $SU(1,1)$). *Any irreducible unitary representation of* $SU(1,1)$ *is equivalent to one of the following:*

1. (*Trivial representation*) 1.
2. (*Discrete series representations*) *For* $p \in \mathbb{Z}_\epsilon$ *and* $p \geq 2$, $D_p^+ = \chi_{\epsilon,p-1}|V_p$ *and* $D_p^- = \chi_{\epsilon,p-1}|\bar{V}_p$.
3. (*Limits of discrete series representation*) D_1 *and* D_{-1}. *Then* $\chi_{1,0} \simeq D_1 \oplus D_{-1}$.
4. (*Unitary principal series representations*) $\chi_{\epsilon,\mu}$ $(\mu \in \sqrt{-1}\mathbb{R} \setminus \mathbb{Z}_{\epsilon'})$.
5. (*Complementary series representations*) $\chi_{0,\mu}$ $(\mu \in \mathbb{R},\ |\mu| < 1)$.

6. Explicit description of Whittaker functions

6.1. The space $C^\infty_{M_0}(A; V_\chi \otimes W_\tau)$ and the radial part of differential operators. In this subsection without loss of generality we assume that $H_\xi = \mathrm{diag}(c_1, c_2)$, i.e. $\gamma = 0$.

The generalized Whittaker function $\Phi_{\pi,\tau}$ for π with K-type τ^* is an element in $C^\infty_{\eta,\tau}(R\backslash G/K)$. We realize this function as a element in a more tractable function space on A by using an decomposition $G = RAK$ of G and standard bases of V_χ and W_τ.

Let us define the space

$$C^{\infty}_{M_0}(A; V_{\chi} \otimes W_{\tau}) = \{ f : A \overset{C^{\infty}}{\longrightarrow} V_{\chi} \otimes W_{\tau} \,|f \text{ satisfies } (\natural) \}. \tag{21}$$

$$(\natural) : f(mam^{-1}) = \eta(m) \otimes \tau(m) f(a) = f(a) \quad \forall m \in M_0. \tag{22}$$

Here, $M_0 = \exp(\mathbb{R} I_0)$, $I_0 = \sqrt{-1}\mathrm{diag}(1, -1, 1, -1) = X_1$. (See Lemma 3.2 on the basis of $\mathfrak{su}(\xi)$.)

Then we have the following.

LEMMA 6.1. *The restriction map* res.A :

$$C^{\infty}_{\eta,\tau}(R\backslash G/K) \ni \Phi_{\pi,\tau}(g) \mapsto \Phi_{\pi,\tau}|_A(a) \in C^{\infty}_{M_0}(A; V_{\chi} \otimes W_{\tau})$$

is a linear injection.

PROOF. Use the fact that $M_0 = SU(\xi) \cap K$ and $M_0 \subset Z_K(A)$ the centralizer of A in K. □

Hereafter we sometimes write $C^{\infty}_{M_0}(A) = C^{\infty}_{M_0}(A; V_{\chi} \otimes W_{\tau})$ and $\Phi_{\pi,\tau}(a) = \Phi_{\pi,\tau}|_A(a)$ for simplicity.

Let us write an element in $C^{\infty}_{M_0}(A)$ more concretely. Recall that $V_{\chi} \otimes W_{\tau}$ are spanned by standard bases.

(a) ξ is definite case

$$V_{\chi} = \bigoplus_{j=0}^{d} \mathbb{C} v_j, \qquad W_{\tau} = \bigoplus_{\substack{0 \le k \le r \\ 0 \le l \le s}} \mathbb{C} f_{kl},$$

for representations $\chi = \mathrm{Sym}^d(\mathbb{C}^2)$ and $\tau = [r, s; u]$.

Hence, an element $\Phi_{\pi,\tau}(a)$ in $C^{\infty}_{M_0}(A)$ can be expressed as

$$\Phi_{\pi,\tau}(a) = \sum_{j=0}^{d} \sum_{\substack{0 \le k \le r \\ 0 \le l \le s}} c_{jkl}(a) \, (v_j \otimes f_{kl})$$

with $(d+1)(r+1)(s+1)$ C^{∞}-functions $c_{jkl}(a)$ on $A \, (\simeq \mathbb{R}^2_{>0})$.

Because $\Phi_{\pi,\tau}(a)$ satisfies the condition $(\natural)$, each function $c_{jkl}(a)$ satisfies that

$$c_{jkl}(a) = e^{\sqrt{-1}(2j-d)\theta} \cdot e^{\sqrt{-1}(2k+2l-r-s)\theta} c_{jkl}(a) \quad (\forall \theta \in \mathbb{R}).$$

Hence, if $c_{jkl}(a)$ is not identically zero then indices j, k, l must satisfy

$$j + k + l = \frac{d + r + s}{2}.$$

In particular if $\frac{d+r+s}{2} \notin \mathbb{Z}$ then $C^{\infty}_{M_0}(A)$ is zero. Therefore we assume that $\frac{j+k+l}{2} \in \mathbb{Z}$ hereafter.

The following case is typical.

LEMMA 6.2. *Assume that* $d + r + s \in 2\mathbb{Z}$ *and* $d \geq r + s$. *Then* $\Phi_{\pi,\tau}(a)$ *can be expressed as*

$$\Phi_{\pi,\tau}(a) = \sum_{\substack{0 \leq k \leq r \\ 0 \leq l \leq s}} c_{kl}(a)\,(v_{\frac{d+r+s}{2}-k-l} \otimes f_{kl})$$

with $(r+1)(s+1)$ C^∞*-functions* $c_{kl}(a)$ *on* A.

(b) ξ is indefinite case

$$V_\chi = \bigoplus_{j \in \mathbb{Z}_\epsilon} \mathbb{C}u_j, \qquad W_\tau = \bigoplus_{\substack{0 \leq k \leq r \\ 0 \leq l \leq s}} \mathbb{C}f_{kl},$$

for representations $\chi = \chi_{\epsilon,\mu}$: non unitary principal series representation and $\tau = [r, s; u]$.

Hence, an element $\Phi_{\pi,\tau}(a)$ in $C^\infty_{M_0}(A)$ can be expressed as

$$\Phi_{\pi,\tau}(a) = \sum_{j \in \mathbb{Z}_\epsilon} \sum_{\substack{0 \leq k \leq r \\ 0 \leq l \leq s}} b_{jkl}(a)\,(u_j \otimes f_{kl})$$

with C^∞-functions $b_{jkl}(a)$ on $A\,(\simeq \mathbb{R}^2_{>0})$.

Because $\Phi_{\pi,\tau}(a)$ satisfies the condition ($\natural$), each function $b_{jkl}(a)$ satisfies that

$$b_{jkl}(a) = e^{\sqrt{-1}j\theta} \cdot e^{\sqrt{-1}(2k+2l-r-s)\theta} b_{jkl}(a) \quad (\forall \theta \in \mathbb{R}).$$

Hence, if $b_{jkl}(a)$ is not identically zero then indices j, k, l must satisfy

$$j + 2k + 2l = r + s.$$

In particular if ϵ is not congruent to $r + s$ modulo 2, then $C^\infty_{M_0}(A)$ is trivial. Therefore we assume that $\epsilon \equiv r + s \pmod 2$

The following case is typical.

LEMMA 6.3. *Assume that* $\epsilon \equiv r+s \pmod 2$ *and that* χ *is unitary principal series representation or complementary series representation. Then* $\Phi_{\pi,\tau}(a)$ *can be expressed as*

$$\Phi_{\pi,\tau}(a) = \sum_{\substack{0 \leq k \leq r \\ 0 \leq l \leq s}} b_{kl}(a)\,(u_{r+s-2k-2l} \otimes f_{kl})$$

with $(r+1)(s+1)$ C^∞*-functions* $b_{kl}(a)$ *on* A.

Let (τ_i, W_{τ_i}) $(i = 1, 2)$ be K-modules and $D\colon C^\infty_{\eta,\tau_1}(R\backslash G/K) \to C^\infty_{\eta,\tau_2}(R\backslash G/K)$, a linear mapping. Set

$$\mathrm{Rad}(D) = \mathrm{res}.A_{\tau_2} \circ D \circ \mathrm{res}.A_{\tau_1}^{-1} : C^\infty_{M_0}(A; V_\chi \otimes W_{\tau_1}) \to C^\infty_{M_0}(A; V_\chi \otimes W_{\tau_2}). \tag{23}$$

We call this linear map $\mathrm{Rad}(D)$ the radial part of D.

For differential operators which characterize generalized Whittaker functions $\Phi_{\pi,\tau}(g)$, we explicitly write down the radial part of such operators which act on the space $C^\infty_{M_0}(A; V_\chi \otimes W_\tau)$.

6.2. The Schmid operator. We introduce the first order gradient type differential operator which is useful for characterizing generalized Whittaker functions. Let (τ, W_τ) be a finite-dimensional irreducible representation of K and $\{X_j\}_{j=1}^8$ an orthonormal basis of $\mathfrak{p}$ with respect to the Killing form on $\mathfrak{g}$. We write the adjoint representation of K on $\mathfrak{p}_\mathbb{C}$ by Ad. Then using the above basis the Schmid operator ∇ is defined by

$$\nabla : C^\infty_\tau(G/K) \ni F \mapsto \sum_{j=0}^{8} X_j.F(\cdot) \otimes X_j \in C^\infty_{\tau\otimes \mathrm{Ad}}(G/K). \qquad (24)$$

Here, right differentiation of the function F by $X \in \mathfrak{g}$ is denoted by $X.F(g) = \frac{d}{dt}F(g \exp tX)|_{t=0}$. This K-equivariant differential operator ∇ is well-defined and independent of the choice of the orthonormal basis. Let us take an orthogonal basis of $\mathfrak{p}$ as

$$\{X_{ij} + X_{ji}, \sqrt{-1}(X_{ij} - X_{ji})\}_{i=1,2,j=3,4}, \qquad (25)$$

where X_{ij} is the matrix element as is defined before. Then using this basis, the Schmid operator ∇ can be decomposed into two terms corresponding to the decomposition $\mathfrak{p}_\mathbb{C} = \mathfrak{p}_+ + \mathfrak{p}_-$.

We define the $\pm$-part of the Schmid operator,

$$\nabla^\pm : C^\infty_\tau(G/K) \to C^\infty_{\tau\otimes \mathrm{Ad}_\pm}(G/K)$$

by

$$\nabla^+ F = \sum_{i=1,2,j=3,4} X_{ji}.F(\cdot) \otimes X_{ij} \qquad (26)$$

$$\nabla^- F = \sum_{i=1,2,j=3,4} X_{ij}.F(\cdot) \otimes X_{ji}. \qquad (27)$$

We can easily check that $\nabla^+ + \nabla^-$ is a constant multiple of ∇ and denote by $\mathrm{Ad}^\pm$ the irreducible components of Ad corresponding to $\mathfrak{p}_\pm$ respectively.

6.3. Radial part of the Schmid operator.

PROPOSITION 6.4. *For the nondegenerate character ξ of N_S, assume that c_2 (or c_1) $\neq 0$. We put $\partial_i = a_i \frac{\partial}{\partial a_i}$ $(i = 1, 2)$ the Euler operator, $\triangle = (c_1 a_1^2 - c_2 a_2^2)$ and $X_5, X_6 \in \mathfrak{su}(\xi)$ are defined in (11). Then the radial parts of the Schmid operators:*

$$\mathrm{Rad}(\nabla^\pm) \colon C^\infty_{M_0}(A; V_\chi \otimes W_\tau) \to C^\infty_{M_0}(A; V_\chi \otimes (W_\tau \otimes \mathfrak{p}_\pm))$$

are given as follows.

$$
\begin{aligned}
&\mathrm{Rad}(\nabla^+)F(a)\\
&= \left[\frac{1}{2}\partial_1 - 2\pi c_1 a_1^2 - \frac{\sqrt{-1}}{2}(\tau \otimes \mathrm{Ad}^+)(H_{13}) - 3 + \frac{2c_1 a_1^2}{\triangle}\right](F(a) \otimes X_{13})\\
&+ \left[\frac{1}{2}\frac{a_1 a_2}{\triangle}\bar{\gamma}(\partial_1 - \partial_2) - 2\pi\sqrt{-1}\gamma a_1 a_2 + \frac{1}{2}\frac{a_1 a_2}{\triangle}\chi(X_5 + \sqrt{-1}X_6)\right.\\
&\quad \left. + (\tau \otimes \mathrm{Ad}^+)(e_+^1) - \frac{c_1 a_1^2}{\triangle}(\tau \otimes \mathrm{Ad}^+)(e_+^1 + e_+^2)\right](F(a) \otimes X_{23})\\
&+ \left[\frac{1}{2}\frac{a_1 a_2}{\triangle}\gamma(\partial_1 - \partial_2) + 2\pi\sqrt{-1}\bar{\gamma} a_1 a_2 + \frac{1}{2}\frac{a_1 a_2}{\triangle}\chi(X_5 - \sqrt{-1}X_6)\right.\\
&\quad \left. - (\tau \otimes \mathrm{Ad}^+)(e_-^2) + \frac{c_1 a_1^2}{\triangle}(\tau \otimes \mathrm{Ad}^+)(e_-^1 + e_-^2)\right](F(a) \otimes X_{14})\\
&+ \left[\frac{1}{2}\partial_2 - 2\pi c_2 a_2^2 - \frac{\sqrt{-1}}{2}(\tau \otimes \mathrm{Ad}^+)(H_{24}) - 1 - \frac{2c_1 a_1^2}{\triangle}\right](F(a) \otimes X_{24}),
\end{aligned}
$$

$$
\begin{aligned}
&\mathrm{Rad}(\nabla^-)F(a)\\
&= \left[\frac{1}{2}\partial_1 + 2\pi c_1 a_1^2 + \frac{\sqrt{-1}}{2}(\tau \otimes \mathrm{Ad}^-)(H_{13}) - 3 + \frac{2c_1 a_1^2}{\triangle}\right](F(a) \otimes X_{31})\\
&+ \left[\frac{1}{2}\frac{a_1 a_2}{\triangle}\bar{\gamma}(\partial_1 - \partial_2) + 2\pi\sqrt{-1}\gamma a_1 a_2 + \frac{1}{2}\frac{a_1 a_2}{\triangle}\chi(X_5 + \sqrt{-1}X_6)\right.\\
&\quad \left. + (\tau \otimes \mathrm{Ad}^-)(e_+^2) - \frac{c_1 a_1^2}{\triangle}(\tau \otimes \mathrm{Ad}^-)(e_+^1 + e_+^2)\right](F(a) \otimes X_{41})\\
&+ \left[\frac{1}{2}\frac{a_1 a_2}{\triangle}\gamma(\partial_1 - \partial_2) - 2\pi\sqrt{-1}\bar{\gamma} a_1 a_2 + \frac{1}{2}\frac{a_1 a_2}{\triangle}\chi(X_5 - \sqrt{-1}X_6)\right.\\
&\quad \left. - (\tau \otimes \mathrm{Ad}^-)(e_-^1) + \frac{c_1 a_1^2}{\triangle}(\tau \otimes \mathrm{Ad}^-)(e_-^1 + e_-^2)\right](F(a) \otimes X_{32})\\
&+ \left[\frac{1}{2}\partial_2 + 2\pi c_2 a_2^2 + \frac{\sqrt{-1}}{2}(\tau \otimes \mathrm{Ad}^-)(H_{24}) - 1 - \frac{2c_1 a_1^2}{\triangle}\right](F(a) \otimes X_{42}).
\end{aligned}
$$

PROOF. We only treat the case of ∇^+, the ∇^- case is quite similar. By definition,

$$
\begin{aligned}
\nabla^+ F = {} & X_{31}.F \otimes X_{13} + X_{32}.F \otimes X_{23}\\
& + X_{41}.F \otimes X_{14} + X_{42}.F \otimes X_{24}. \qquad (28)
\end{aligned}
$$

We use the Iwasawa decomposition of X_{31}, X_{32}, X_{41} and X_{42}, (see Lemma 3.5), and use the expression of E_5, E_6 in Lemma 3.3. Then we have the decomposition of X_{31}, X_{32}, X_{41} and X_{42} along the decomposition: $\mathfrak{g} = \mathrm{Ad}(a^{-1})(\mathfrak{su}(\xi) + \mathfrak{n}_S) + \mathfrak{a} + \mathfrak{k}$. Then we have

$$\begin{aligned}
&2X_{31}.F \otimes X_{13} \\
&= (-\sqrt{-1}E_1 + H_1 + \sqrt{-1}H_{13}).F \otimes X_{13} \\
&= [\partial_1 - \sqrt{-1}a_1^2\xi(E_1) - \sqrt{-1}(\tau \otimes \mathrm{Ad}^+)(H_{13}) - 2]\, F \otimes X_{13},
\end{aligned} \tag{29}$$

$$\begin{aligned}
&2X_{32}.F \otimes X_{23} \\
&= (E_3 - \sqrt{-1}E_4 - 2e_+^1 + E_5 - \sqrt{-1}E_6).F \otimes X_{23} \\
&= (E_3 - \sqrt{-1}E_4 - 2e_+^1).F \otimes X_{23} \\
&\quad + \frac{a_1a_2}{\triangle}\Big\{\mathrm{Ad}(a^{-1})(X_5 + \sqrt{-1}X_6) + c_1(\frac{a_1}{a_2})(Y_5 - \sqrt{-1}Y_6) \\
&\qquad\qquad + \bar{\gamma}(H_1 - H_2)\Big\}.F \otimes X_{23} \\
&= \Big[a_1a_2\xi(E_3 - \sqrt{-1}E_4) + 2(\tau \otimes \mathrm{Ad}^+)(e_+^1) \\
&\quad + \frac{a_1a_2}{\triangle}\chi(X_5 + \sqrt{-1}X_6) - 2\frac{c_1a_1^2}{\triangle}(\tau \otimes \mathrm{Ad}^+)(e_+^1 + e_+^2) \\
&\quad + \frac{a_1a_2}{\triangle}\bar{\gamma}(\partial_1 - \partial_2)\Big]\, F \otimes X_{23} \\
&\quad -2F \otimes X_{13} + 2\frac{c_1a_1^2}{\triangle}(F \otimes X_{13} - F \otimes X_{24}),
\end{aligned} \tag{30}$$

$$\begin{aligned}
&2X_{41}.F \otimes X_{14} \\
&= (-E_3 - \sqrt{-1}E_4 + 2e_-^2 + E_5 + \sqrt{-1}E_6).F \otimes X_{14} \\
&= (-E_3 - \sqrt{-1}E_4 + 2e_-^2).F \otimes X_{14} \\
&\quad + \frac{a_1a_2}{\triangle}\Big\{\mathrm{Ad}(a^{-1})(X_5 - \sqrt{-1}X_6) + c_1(\frac{a_1}{a_2})(Y_5 + \sqrt{-1}Y_6) \\
&\qquad\qquad + \gamma(H_1 - H_2)\Big\}.F \otimes X_{14} \\
&= \Big[-a_1a_2\xi(E_3 + \sqrt{-1}E_4) - 2(\tau \otimes \mathrm{Ad}^+)(e_-^2) \\
&\quad + \frac{a_1a_2}{\triangle}\chi(X_5 - \sqrt{-1}X_6) + 2\frac{c_1a_1^2}{\triangle}(\tau \otimes \mathrm{Ad}^+)(e_-^1 + e_-^2) \\
&\quad + \frac{a_1a_2}{\triangle}\gamma(\partial_1 - \partial_2)\Big]\, F \otimes X_{14} \\
&\quad -2F \otimes X_{13} + 2\frac{c_1a_1^2}{\triangle}(F \otimes X_{13} - F \otimes X_{24}),
\end{aligned} \tag{31}$$

$$\begin{aligned}
&2X_{42}.F \otimes X_{24} \\
&= (-\sqrt{-1}E_2 + H_2 + \sqrt{-1}H_{24}).F \otimes X_{24} \\
&= [\partial_2 - \sqrt{-1}a_2^2\xi(E_2) - \sqrt{-1}(\tau \otimes \mathrm{Ad}^+)(H_{24}) - 2]\, F \otimes X_{24}. \quad (32)
\end{aligned}$$

In the above calculation, we used the following formula: (similar as in [**Kn**, pp. 215-218]

$$\begin{aligned}
H_i.F(a) &= \partial_i F(a) \quad (i = 1, 2), \\
\mathrm{Ad}(a^{-1})(X).F(a) &= \chi(X)\, F(a) \qquad \text{for} \quad X \in \mathfrak{su}(\xi)_{\mathbb{C}}, \\
\mathrm{Ad}(a^{-1})(E).F(a) &= \xi(E)\, F(a) \qquad \text{for} \quad E \in \mathfrak{n}_{S,\mathbb{C}}, \\
e.F(a) &= -\tau(e)\, F(a) \qquad \text{for} \quad e \in \mathfrak{k}_{\mathbb{C}},
\end{aligned}$$

and

$$\begin{aligned}
\sqrt{-1}H_{13}.F \otimes X_{13} &= -\sqrt{-1}(\tau \otimes \mathrm{Ad}^+)(H_{13})(F \otimes X_{13}) - 2F \otimes X_{13}, \\
-e_+^1.F \otimes X_{23} &= \quad (\tau \otimes \mathrm{Ad}^+)(e_+^1)(F \otimes X_{23}) - F \otimes X_{13}, \\
-e_+^2.F \otimes X_{23} &= \quad (\tau \otimes \mathrm{Ad}^+)(e_+^2)(F \otimes X_{23}) + F \otimes X_{24}, \\
e_-^2.F \otimes X_{14} &= -(\tau \otimes \mathrm{Ad}^+)(e_-^2)(F \otimes X_{14}) - F \otimes X_{13}, \\
e_-^1.F \otimes X_{14} &= -(\tau \otimes \mathrm{Ad}^+)(e_-^1)(F \otimes X_{14}) + F \otimes X_{24}, \\
\sqrt{-1}H_{24}.F \otimes X_{24} &= -\sqrt{-1}(\tau \otimes \mathrm{Ad}^+)(H_{24})(F \otimes X_{24}) - 2F \otimes X_{24}.
\end{aligned}$$

And concerning on the infinitesimal representation of ξ:

$$\begin{aligned}
\xi(E_1) &= -4\pi\sqrt{-1}c_1, & \xi(E_2) &= -4\pi\sqrt{-1}c_2, \\
\xi(E_3) &= -4\pi\sqrt{-1}\mathrm{Re}(\gamma), & \xi(E_4) &= 4\pi\sqrt{-1}\mathrm{Im}(\gamma).
\end{aligned}$$

We complete the proof. □

6.4. Radial part of the Casimir operator. The Casimir operator $\Omega \in Z(\mathfrak{g}_{\mathbb{C}})$ is given by,

$$\Omega = H_1^2 + H_2^2 + \frac{1}{2}I_0^2 - \frac{1}{2}\sum_{j=1,2}(E_j{}^tE_j + {}^tE_jE_j) + \sum_{j=3}^{6}(E_j{}^t\bar{E}_j + {}^t\bar{E}_jE_j). \quad (33)$$

PROPOSITION 6.5. *Assume that* $\det H_\xi \neq 0$ *and* $\gamma = 0$. *Then the radial part of the Casimir operator is given as follows. Here, we use the same notation in Proposition 6.4 and* $X_5, X_6 \in \mathfrak{su}(\xi)$ *are defined in (11).*

$$\begin{aligned}
\mathrm{Rad}(\Omega) &= \partial_1^2 + \partial_2^2 - 2(\partial_1 + \partial_2) + 4\frac{c_2a_2^2}{\triangle}\partial_1 - 4\frac{c_1a_1^2}{\triangle}\partial_2 \\
&\quad - 16\pi^2(c_1^2a_1^4 + c_2^2a_2^4) - 8\pi c_1a_1^2\tau(\sqrt{-1}H_{13}) - 8\pi c_2a_2^2\tau(\sqrt{-1}H_{24})
\end{aligned}$$

$$- 4\frac{a_1a_2}{\triangle}\frac{c_1a_1^2+c_2a_2^2}{\triangle}\{\chi(X_5+\sqrt{-1}X_6)\tau(e_-^{12}) \\ -\chi(X_5-\sqrt{-1}X_6)\tau(e_+^{12})\} \\ - 4c_1c_2\left(\frac{a_1a_2}{\triangle}\right)^2\{\tau(e_+^{12})\tau(e_-^{12})+\tau(e_-^{12})\tau(e_+^{12})\} \\ + 2\left(\frac{a_1a_2}{\triangle}\right)^2(\chi(X_5)^2+\chi(X_6)^2)+\frac{1}{2}\tau(h^1+h^2)^2$$

Here, $e_+^{12}=e_+^1+e_+^2$ *and* $e_-^{12}=e_-^1+e_-^2$ *are elements in* $\mathfrak{k}_{\mathbb{C}}$.

PROOF. We firstly define four elements Ω_i $(i=1,...,4)$ of the universal enveloping algebra of $\mathfrak{g}_{\mathbb{C}}$. They are given by

$$\Omega_1 = H_1^2+H_2^2-6H_1-2H_2+\frac{1}{2}I_0^2+(E_1^2+E_2^2+2E_3^2+2E_4^2) \\ -2E_1H_{13}-2E_2H_{24},$$
$$\Omega_2 = -2E_3(e_+^1-e_-^1-e_+^2+e_-^2)-2E_4\sqrt{-1}(e_+^1+e_-^1+e_+^2+e_-^2),$$
$$\Omega_3 = -2E_5(e_+^1-e_-^1+e_+^2-e_-^2)-2E_6\sqrt{-1}(e_+^1+e_-^1+e_+^2+e_-^2),$$
$$\Omega_4 = 2E_5^2+2E_6^2.$$

Then we can easily check that

$$\Omega=\Omega_1+\Omega_2+\Omega_3+\Omega_4.$$

So we calculate each $\mathrm{Rad}(\Omega_i)F(a)$ for $F(a)\in C_{M_0}^\infty(A;V_\chi\otimes W_\tau)$. Then we have the following formula similar to the case of the Schmid operator. Note that the assumption $\gamma=0$ implies that $\xi(E_3)=\xi(E_4)=0$.

$$\mathrm{Rad}(\Omega_1)F(a) \\ = [\,\partial_1^2+\partial^2-6\partial_1-2\partial_2+(a_1^2\xi(E_1))^2+(a_2^2\xi(E_2))^2 \\ +2a_1^2\xi(E_1)\tau(H_{13})+2a_2^2\xi(E_2)\tau(H_{24})\,]\,F(a),$$
$$\mathrm{Rad}(\Omega_2)F(a)=0,$$
$$\mathrm{Rad}(\Omega_3)F(a) \\ = 2[\,(E_5+\sqrt{-1}E_6)\tau(e_+^1)-(E_5-\sqrt{-1}E_6)\tau(e_-^1) \\ (E_5+\sqrt{-1}E_6)\tau(e_+^2)-(E_5-\sqrt{-1}E_6)\tau(e_-^2)\,]\,F(a) \\ = \frac{2a_1a_2}{\triangle}\Big[\chi(X_5-\sqrt{-1}X_6)\tau(e_+^{12})+2c_1(\frac{a_1}{a_2})\tau(e_-^{12})\tau(e_+^{12}) \\ -\chi(X_5+\sqrt{-1}X_6)\tau(e_-^{12})+2c_1(\frac{a_1}{a_2})\tau(e_+^{12})\tau(e_-^{12})\Big]\,F(a).$$

To get $\mathrm{Rad}(\Omega_4)$, we calculate $2E_5^2.F(a)$, $2E_6^2.F(a)$.

$2E_5^2.F(a)$

$$= 2\left(\frac{a_1a_2}{\triangle}\right)^2 \left[\mathrm{Ad}(a^{-1})X_5 + c_1(\frac{a_1}{a_2})(Y_5)\right]^2 .F(a)$$

$$= 2\left(\frac{a_1a_2}{\triangle}\right)^2 \left[(\mathrm{Ad}(a^{-1})X_5)^2 + c_1^2(\frac{a_1}{a_2})^2 Y_5^2 + 2c_1(\frac{a_1}{a_2})\mathrm{Ad}(a^{-1})X_5Y_5 \right.$$
$$\left. + c_1\frac{\triangle}{a_2^2}(H_1 - H_2)\right] F(a).$$

Here, we used the equality:

$$[Y_5, \mathrm{Ad}(a^{-1})(X_5)]$$
$$= \left[Y_5, \frac{\triangle}{a_1a_2}E_5 + c_1(\frac{a_1}{a_2})Y_5\right] = \frac{\triangle}{a_1a_2}[E_{-5}, E_5] = \frac{\triangle}{a_1a_2}(H_1 - H_2).$$

We have similarly,

$$2E_6^2.F(a)$$
$$= 2\left(\frac{a_1a_2}{\triangle}\right)^2 \left[(\mathrm{Ad}(a^{-1})X_6)^2 + c_1^2(\frac{a_1}{a_2})^2 Y_6^2 - 2c_1(\frac{a_1}{a_2})\mathrm{Ad}(a^{-1})X_6Y_6 \right.$$
$$\left. + c_1\frac{\triangle}{a_2^2}(H_1 - H_2)\right] F(a).$$

Thus we have

$$\mathrm{Rad}(\Omega_4)F(a)$$
$$= 2\left(\frac{a_1a_2}{\triangle}\right)^2 \left[\chi(X_5)^2 + \chi(X_6)^2 + c_1^2(\frac{a_1}{a_2})^2\{\tau(Y_5)^2 + \tau(Y_6)^2\} \right.$$
$$\left. + 2c_1(\frac{a_1}{a_2})\{\chi(X_6)\tau(Y_6) - \chi(X_5)\tau(Y_5)\} + \frac{2c_1\triangle}{a_2^2}(\partial_1 - \partial_2)\right] F(a)$$

So we have $\mathrm{Rad}(\Omega) = \sum_{i=1}^4 \mathrm{Rad}(\Omega_i)$. □

7. Radial part of the shift operators

Let $\Phi_{\pi,\tau}(a) = \sum_{j=0}^d \sum_{\substack{0\le k\le r \\ 0\le l\le s}} c_{jkl}(a)(v_j \otimes f_{kl}) \in C_{M_0}^\infty(A; V_\chi \otimes W_\tau)$. Here, $\eta = \chi \otimes \xi$, χ is an irreducible finite-dimensional representation of $SU(\xi) \simeq SU(2)$ and ξ is a non-trivial unitary character of N_S determined by a positive definite matrix $H_\xi = \begin{pmatrix} c_1 & \\ & c_2 \end{pmatrix}$. We assume that $\dim \chi = d + 1$ and $d + r + s \in 2\mathbb{Z}$.

We calculate $P \circ \mathrm{Rad}(\nabla)c_{jkl}(a)(v_j \otimes f_{kl})$ in case of the projector $P = P^{(-,-)}$, $P^{(+,-)}$, $P^{(-,+)}$, $\bar{P}^{(-,-)}$, $\bar{P}^{(+,-)}$ and $\bar{P}^{(-,+)}$ to get the differential equations which characterize $\Phi_{\pi,\tau}(a)$.

We put

$$b_0 = \frac{1}{2}(r+s+u), \qquad b_1 = \frac{1}{2}(-r+s+u),$$
$$b_2 = \frac{1}{2}(r-s+u), \qquad b_3 = \frac{1}{2}(-r-s+u), \tag{34}$$

which are integers since $r+s+u \in 2\mathbb{Z}$.

7.1. $P^{(-,-)} \circ \mathrm{Rad}(\nabla^+) c_{jkl}(a)(v_j \otimes f_{kl})$**.**

LEMMA 7.1. *If we write*

$$\begin{aligned} &P^{(-,-)} \circ \mathrm{Rad}(\nabla^+) c_{jkl}(a)(v_j \otimes f_{kl}) \\ &= A^{(-,-)}_{jkl}(a)(v_{j+1} \otimes f_{k-1,l-1}) + B^{(-,-)}_{jkl}(a)(v_{j-1} \otimes f_{k,l}) \\ &\quad + C^{(-,-)}_{jkl}(a)(v_j \otimes f_{k,l-1}) + D^{(-,-)}_{jkl}(a)(v_j \otimes f_{k-1,l}), \end{aligned} \tag{35}$$

then the above four functions are given as follows:

$$A^{(-,-)}_{jkl}(a) = -\sqrt{c_1 c_2}\frac{a_1 a_2}{\triangle}(d-j)kl\, c_{jkl}(a),$$

$$B^{(-,-)}_{jkl}(a) = \sqrt{c_1 c_2}\frac{a_1 a_2}{\triangle} j(k-r)(s-l)\, c_{jkl}(a),$$

$$\begin{aligned} C^{(-,-)}_{jkl}(a) = \frac{1}{2}(k-r)l\Big[&\partial_1 - 4\pi c_1 a_1^2 + (b_1 - k + l - 2s - 4) \\ &+(2+k+s-l)\frac{2c_1 a_1^2}{\triangle}\Big] c_{jkl}(a), \end{aligned}$$

$$\begin{aligned} D^{(-,-)}_{jkl}(a) = \frac{1}{2}k(s-l)\Big[&\partial_2 - 4\pi c_2 a_2^2 + (b_2 + k - 2r - l - 4) \\ &-(2+r-k+l+)\frac{2c_2 a_2^2}{\triangle}\Big] c_{jkl}(a). \end{aligned}$$

Here, we put $\partial_i = a_i \frac{\partial}{\partial a_i}$ *: the Euler operator* $(i=1,2)$ *and* $\triangle = (c_1 a_1^2 - c_2 a_2^2)$*.*

PROOF. We first note that:

$$\tau \otimes \mathrm{Ad}^+ = [r,s;u] \otimes [1,1;2] \overset{P^{(-,-)}}{\longrightarrow} [r-1, s-1; u+2],$$

and set $\tau' = [r-1, s-1; u+2]$.

$$\begin{aligned} &P^{(-,-)} \circ \mathrm{Rad}(\nabla^+) c_{jkl}(a)(v_j \otimes f_{kl}) \\ &= \Big[\frac{1}{2}\partial_1 - 2\pi c_1 a_1^2 + \frac{1}{2}\tau'(-\sqrt{-1}H_{13}) \\ &\qquad -3 + \frac{2c_1 a_1^2}{\triangle}\Big] (c_{jkl}(a) v_j \otimes [(k-r) l f_{k,l-1}]) \end{aligned}$$

$$+\left[\frac{1}{2}\frac{a_1a_2}{\triangle}\chi(X_5+\sqrt{-1}X_6)+\tau'(e_+^1)\right.$$
$$\left.-\frac{c_1a_1^2}{\triangle}\tau'(e_+^1+e_+^2)\right](c_{jkl}(a)v_j\otimes[klf_{k-1,l-1}])$$
$$+\left[\frac{1}{2}\frac{a_1a_2}{\triangle}\chi(X_5-\sqrt{-1}X_6)-\tau'(e_-^2)\right.$$
$$\left.+\frac{c_1a_1^2}{\triangle}\tau'(e_-^1+e_-^2)\right](c_{jkl}(a)v_j\otimes[(k-r)(s-l)f_{k,l}])$$
$$+\left[\frac{1}{2}\partial_2-2\pi c_2a_2^2+\frac{1}{2}\tau'(-\sqrt{-1}H_{24})\right.$$
$$\left.-1-\frac{2c_1a_1^2}{\triangle}\right](c_{jkl}(a)v_j\otimes[k(s-l)f_{k-1,l}])$$

Here, we use the fact (from Lemma 4.4):

$$\begin{aligned}
P^{(-,-)}(f_{kl}\otimes X_{13})&=(k-r)lf_{k,l-1},\\
P^{(-,-)}(f_{kl}\otimes X_{23})&=klf_{k-1,l-1},\\
P^{(-,-)}(f_{kl}\otimes X_{14})&=(k-r)(s-l)f_{k,l},\\
P^{(-,-)}(f_{kl}\otimes X_{24})&=k(s-l)f_{k-1,l}.
\end{aligned}$$

Next we use the formulas from Lemmas 4.1 and 5.1:

$$\begin{aligned}
&\tau'(-\sqrt{-1}H_{13})(f_{k,l-1})=(b_1+k-l+2)f_{k,l-1},\\
&\tau'(e_+^1)(f_{k-1,l-1})=(r-k)f_{k,l-1},\quad \tau'(e_+^2)(f_{k-1,l-1})=(s-l)f_{k-1,l},\\
&\tau'(e_-^2)(f_{k,l})=lf_{k,l-1},\qquad \tau'(e_-^1)(f_{k,l})=kf_{k-1,l},\\
&\tau'(-\sqrt{-1}H_{24})(f_{k-1,l})=(b_2-k+l+2)f_{k-1,l}.
\end{aligned}$$

and

$$\begin{aligned}
\frac{1}{2}\frac{a_1a_2}{\triangle}\chi(X_5+\sqrt{-1}X_6)v_j&=-\sqrt{c_1c_2}\frac{a_1a_2}{\triangle}(d-j)v_{j+1},\\
\frac{1}{2}\frac{a_1a_2}{\triangle}\chi(X_5-\sqrt{-1}X_6)v_j&=\sqrt{c_1c_2}\frac{a_1a_2}{\triangle}jv_{j-1}.
\end{aligned}$$

Then we have:

$$\begin{aligned}
&P^{(-,-)}\circ\mathrm{Rad}(\nabla^+)c_{jkl}(a)(v_j\otimes f_{kl})\\
&=(k-r)l\left[\frac{1}{2}\partial_1-2\pi c_1a_1^2+\frac{1}{2}(b_1+k-l-4)\right.\\
&\qquad\left.+\frac{2c_1a_1^2}{\triangle}\right](c_{jkl}(a)v_j\otimes f_{k,l-1})
\end{aligned}$$

$$+kl\left[-\sqrt{c_1c_2}\frac{a_1a_2}{\triangle}\chi(X_+)+\tau'(e^1_+)\right.$$
$$\left.-\frac{c_1a_1^2}{\triangle}\tau'(e^1_++e^2_+)\right](c_{jkl}(a)v_j\otimes f_{k-1,l-1})$$
$$+(k-r)(s-l)\left[\sqrt{c_1c_2}\frac{a_1a_2}{\triangle}\chi(X_-)-\tau'(e^2_-)\right.$$
$$\left.+\frac{c_1a_1^2}{\triangle}\tau'(e^1_-+e^2_-)\right](c_{jkl}(a)v_j\otimes f_{k,l})$$
$$+k(s-l)\left[\frac{1}{2}\partial_2-2\pi c_2a_2^2+\frac{1}{2}(b_2-k+l)\right.$$
$$\left.-\frac{2c_1a_1^2}{\triangle}\right](c_{jkl}(a)v_j\otimes f_{k-1,l})$$
$$=(k-r)l\left[\frac{1}{2}\partial_1-2\pi c_1a_1^2+\frac{1}{2}(b_1+k-l-4)\right.$$
$$\left.+\frac{2c_1a_1^2}{\triangle}\right]c_{jkl}(a)(v_j\otimes f_{k,l-1})$$
$$-kl\sqrt{c_1c_2}\frac{a_1a_2}{\triangle}(d-j)\,c_{jkl}(a)(v_{j+1}\otimes f_{k-1,l-1})$$
$$+kl(1-\frac{c_1a_1^2}{\triangle})(r-k)\,c_{jkl}(a)(v_j\otimes f_{k,l-1})$$
$$-kl\frac{c_1a_1^2}{\triangle}(s-l)\,c_{jkl}(a)(v_j\otimes f_{k-1,l})$$
$$+(k-r)(s-l)\sqrt{c_1c_2}\frac{a_1a_2}{\triangle}j\,c_{jkl}(a)(v_{j-1}\otimes f_{k,l})$$
$$+(k-r)(s-l)(-1+\frac{c_1a_1^2}{\triangle})l\,c_{jkl}(a)(v_j\otimes f_{k,l-1})$$
$$+(k-r)(s-l)\frac{c_1a_1^2}{\triangle}k\,c_{jkl}(a)(v_j\otimes f_{k-1,l+1})$$
$$+k(s-l)\left[\frac{1}{2}\partial_2-2\pi c_2a_2^2+\frac{1}{2}(b_2-k+l)\right.$$
$$\left.-\frac{2c_1a_1^2}{\triangle}\right]c_{jkl}(a)(v_j\otimes f_{k-1,l})$$

Therefore we have :

$$P^{(-,-)}\circ\mathrm{Rad}(\nabla^+)c_{jkl}(a)(v_j\otimes f_{kl})$$
$$=-\sqrt{c_1c_2}\frac{a_1a_2}{\triangle}(d-j)kl\,c_{jkl}(a)(v_{j+1}\otimes f_{k-1,l-1})$$

$$+\sqrt{c_1c_2}\frac{a_1a_2}{\triangle}j(k-r)(s-l)\,c_{jkl}(a)(v_{j-1}\otimes f_{k,l})$$
$$+(k-r)l\left[\frac{1}{2}\partial_1-2\pi c_1a_1^2+\frac{1}{2}(b_1+k-l-4)+\frac{2c_1a_1^2}{\triangle}-k\right.$$
$$\left.+k\frac{c_1a_1^2}{\triangle}-(s-l)+(s-l)\frac{c_1a_1^2}{\triangle}\right]\,c_{jkl}(a)(v_j\otimes f_{k,l-1})$$
$$+k(s-l)\left[\frac{1}{2}\partial_2-2\pi c_2a_2^2+\frac{1}{2}(b_2-k+l)-\frac{2c_1a_1^2}{\triangle}\right.$$
$$\left.-l\frac{c_1a_1^2}{\triangle}+(k-r)\frac{c_1a_1^2}{\triangle}\right]\,c_{jkl}(a)(v_j\otimes f_{k-1,l}).$$

So we get the desired formula. □

7.2. $P^{(-,+)}\circ\mathrm{Rad}(\nabla^+)c_{jkl}(a)(v_j\otimes f_{kl})$.

LEMMA 7.2. *If we write*

$$\begin{aligned}&P^{(-,+)}\circ\mathrm{Rad}(\nabla^+)c_{jkl}(a)(v_j\otimes f_{kl})\\&=A_{jkl}^{(-,+)}(a)(v_{j+1}\otimes f_{k-1,l})+B_{jkl}^{(-,+)}(a)(v_{j-1}\otimes f_{k,l+1})\\&\quad+C_{jkl}^{(-,+)}(a)(v_j\otimes f_{k,l})+D_{jkl}^{(-,+)}(a)(v_j\otimes f_{k-1,l+1}),\end{aligned}\tag{36}$$

then the above four functions are given as follows:

$$A_{jkl}^{(-,+)}(a)=\sqrt{c_1c_2}\frac{a_1a_2}{\triangle}(d-j)k\,c_{jkl}(a),$$
$$B_{jkl}^{(-,+)}(a)=\sqrt{c_1c_2}\frac{a_1a_2}{\triangle}j(k-r)\,c_{jkl}(a),$$
$$\begin{aligned}C_{jkl}^{(-,+)}(a)=\frac{1}{2}(r-k)\Big[&\partial_1-4\pi c_1a_1^2+(b_1-k+l-2)\\&+(1+k-l)\frac{2c_1a_1^2}{\triangle}\Big]c_{jkl}(a),\end{aligned}$$
$$\begin{aligned}D_{jkl}^{(-,+)}(a)=\frac{1}{2}k\Big[&\partial_2-4\pi c_2a_2^2+(b_2+k-l-2r+2s-2)\\&-(1-k+l+r-s)\frac{2c_2a_2^2}{\triangle}\Big]c_{jkl}(a).\end{aligned}$$

Here, we put $\partial_i=a_i\frac{\partial}{\partial a_i}$ *: the Euler operator* $(i=1,2)$ *and* $\triangle=(c_1a_1^2-c_2a_2^2)$.

PROOF. We first note that:

$$\tau\otimes\mathrm{Ad}^+=[r,s;u]\otimes[1,1;2]\xrightarrow{P^{(\ ,\,)}}[r-1,s+1;u+2],$$

and set $\tau'=[r-1,s+1;u+2]$.

$$
\begin{aligned}
&P^{(-,+)} \circ \mathrm{Rad}(\nabla^+) c_{jkl}(a)(v_j \otimes f_{kl}) \\
&= \left[\frac{1}{2}\partial_1 - 2\pi c_1 a_1^2 + \frac{1}{2}\tau'(-\sqrt{-1}H_{13}) \right.\\
&\qquad \left. -3 + \frac{2c_1a_1^2}{\triangle}\right] (c_{jkl}(a)v_j \otimes [(r-k)f_{k,l}]) \\
&\quad + \left[\frac{1}{2}\frac{a_1a_2}{\triangle}\chi(X_5 + \sqrt{-1}X_6) + \tau'(e_+^1) \right.\\
&\qquad \left. -\frac{c_1a_1^2}{\triangle}\tau'(e_+^1 + e_+^2)\right] (c_{jkl}(a)v_j \otimes [-kf_{k-1,l}]) \\
&\quad + \left[\frac{1}{2}\frac{a_1a_2}{\triangle}\chi(X_5 - \sqrt{-1}X_6) - \tau'(e_-^2) \right.\\
&\qquad \left. +\frac{c_1a_1^2}{\triangle}\tau'(e_-^1 + e_-^2)\right] (c_{jkl}(a)v_j \otimes [(k-r)f_{k,l+1}]) \\
&\quad + \left[\frac{1}{2}\partial_2 - 2\pi c_2 a_2^2 + \frac{1}{2}\tau'(-\sqrt{-1}H_{24}) \right.\\
&\qquad \left. -1 - \frac{2c_1a_1^2}{\triangle}\right] (c_{jkl}(a)v_j \otimes [kf_{k-1,l+1}])
\end{aligned}
$$

Here, we use the fact (from Lemma 4.4):

$$
\begin{aligned}
P^{(-,+)}(f_{kl} \otimes X_{13}) &= (r-k)f_{k,l}, & P^{(-,+)}(f_{kl} \otimes X_{23}) &= -kf_{k-1,l}, \\
P^{(-,+)}(f_{kl} \otimes X_{14}) &= (k-r)f_{k,l+1}, & P^{(-,+)}(f_{kl} \otimes X_{24}) &= kf_{k-1,l+1}.
\end{aligned}
$$

Next we use the formulas from Lemmas 4.1 and 5.1:

$$
\begin{aligned}
&\tau'(-\sqrt{-1}H_{13})(f_{k,l}) = (b_1 + k - l + 2)f_{k,l}, \\
&\tau'(e_+^1)(f_{k-1,l}) = (r-k)f_{k,l}, \quad \tau'(e_+^2)(f_{k-1,l}) = (s+1-l)f_{k-1,l+1}, \\
&\tau'(e_-^2)(f_{k,l+1}) = (l+1)f_{k,l}, \qquad \tau'(e_-^1)(f_{k,l+1}) = kf_{k-1,l+1}, \\
&\tau'(-\sqrt{-1}H_{24})(f_{k-1,l+1}) = (b_2 - k + l + 2)f_{k-1,l+1}.
\end{aligned}
$$

and

$$
\begin{aligned}
\frac{1}{2}\frac{a_1a_2}{\triangle}\chi(X_5 + \sqrt{-1}X_6)v_j &= -\sqrt{c_1c_2}\frac{a_1a_2}{\triangle}(d-j)v_{j+1}, \\
\frac{1}{2}\frac{a_1a_2}{\triangle}\chi(X_5 - \sqrt{-1}X_6)v_j &= \sqrt{c_1c_2}\frac{a_1a_2}{\triangle}jv_{j-1}.
\end{aligned}
$$

Then we have:

$$
P^{(-,+)} \circ \mathrm{Rad}(\nabla^+) c_{jkl}(a)(v_j \otimes f_{kl})
$$

$$
\begin{aligned}
&= (r-k)\left[\frac{1}{2}\partial_1 - 2\pi c_1 a_1^2 + \frac{1}{2}(b_1+k-l-4)\right.\\
&\qquad\qquad \left. + \frac{2c_1a_1^2}{\triangle}\right](c_{jkl}(a)v_j \otimes f_{k,l})\\
&\quad -k\left[-\sqrt{c_1c_2}\frac{a_1a_2}{\triangle}\chi(X_+) + \tau'(e_+^1)\right.\\
&\qquad\qquad \left. - \frac{c_1a_1^2}{\triangle}\tau'(e_+^1 + e_+^2)\right](c_{jkl}(a)v_j \otimes f_{k-1,l})\\
&\quad +(k-r)\left[\sqrt{c_1c_2}\frac{a_1a_2}{\triangle}\chi(X_-) - \tau'(e_-^2)\right.\\
&\qquad\qquad \left. + \frac{c_1a_1^2}{\triangle}\tau'(e_-^1 + e_-^2)\right](c_{jkl}(a)v_j \otimes f_{k,l+1})\\
&\quad +k\left[\frac{1}{2}\partial_2 - 2\pi c_2 a_2^2 + \frac{1}{2}(b_2-k+l)\right.\\
&\qquad\qquad \left. - \frac{2c_1a_1^2}{\triangle}\right](c_{jkl}(a)v_j \otimes f_{k-1,l+1})\\
&= (r-k)\left[\frac{1}{2}\partial_1 - 2\pi c_1 a_1^2 + \frac{1}{2}(b_1+k-l-4)\right.\\
&\qquad\qquad \left. + \frac{2c_1a_1^2}{\triangle}\right] c_{jkl}(a)(v_j \otimes f_{k,l})\\
&\quad +k\sqrt{c_1c_2}\frac{a_1a_2}{\triangle}(d-j)\, c_{jkl}(a)(v_{j+1} \otimes f_{k-1,l})\\
&\qquad -k(1-\frac{c_1a_1^2}{\triangle})(r-k)\, c_{jkl}(a)(v_j \otimes f_{k,l})\\
&\qquad +k\frac{c_1a_1^2}{\triangle}(s+1-l)\, c_{jkl}(a)(v_j \otimes f_{k-1,l+1})\\
&\quad +(k-r)\sqrt{c_1c_2}\frac{a_1a_2}{\triangle}j\, c_{jkl}(a)(v_{j-1} \otimes f_{k,l+1})\\
&\qquad +(k-r)(-1+\frac{c_1a_1^2}{\triangle})(l+1)\, c_{jkl}(a)(v_j \otimes f_{k,l})\\
&\qquad +(k-r)\frac{c_1a_1^2}{\triangle}k\, c_{jkl}(a)(v_j \otimes f_{k-1,l+1})\\
&\quad +k\left[\frac{1}{2}\partial_2 - 2\pi c_2 a_2^2 + \frac{1}{2}(b_2-k+l) - \frac{2c_1a_1^2}{\triangle}\right] c_{jkl}(a)(v_j \otimes f_{k-1,l+1})
\end{aligned}
$$

Therefore we have :

$$P^{(-,+)} \circ \mathrm{Rad}(\nabla^+) c_{jkl}(a)(v_j \otimes f_{kl})$$

$$
\begin{aligned}
&= \sqrt{c_1c_2}\frac{a_1a_2}{\triangle}(d-j)k\, c_{jkl}(a)(v_{j+1}\otimes f_{k-1,l}) \\
&+\sqrt{c_1c_2}\frac{a_1a_2}{\triangle}j(k-r)\, c_{jkl}(a)(v_{j-1}\otimes f_{k,l+1}) \\
&+(r-k)\left[\frac{1}{2}\partial_1 - 2\pi c_1a_1^2 + \frac{1}{2}(b_1+k-l-4) + \frac{2c_1a_1^2}{\triangle} - k\right. \\
&\qquad \left. +k\frac{c_1a_1^2}{\triangle} + (l+1) - (l+1)\frac{c_1a_1^2}{\triangle}\right] c_{jkl}(a)(v_j\otimes f_{k,l}) \\
&+k\left[\frac{1}{2}\partial_2 - 2\pi c_2a_2^2 + \frac{1}{2}(b_2-k+l) - \frac{2c_1a_1^2}{\triangle}\right. \\
&\qquad \left.(s+1-l)\frac{c_1a_1^2}{\triangle} + (k-r)\frac{c_1a_1^2}{\triangle}\right] c_{jkl}(a)(v_j\otimes f_{k-1,l+1}).
\end{aligned}
$$

So we get the desired formula. □

7.3. $P^{(+,-)}\circ \mathrm{Rad}(\nabla^+)c_{jkl}(a)(v_j\otimes f_{kl})$.

LEMMA 7.3. *If we write*

$$
\begin{aligned}
&P^{(+,-)}\circ \mathrm{Rad}(\nabla^+)c_{jkl}(a)(v_j\otimes f_{kl}) \\
&= A_{jkl}^{(+,-)}(a)(v_{j+1}\otimes f_{k,l-1}) + B_{jkl}^{(+,-)}(a)(v_{j-1}\otimes f_{k+1,l}) \\
&\quad +C_{jkl}^{(+,-)}(a)(v_j\otimes f_{k+1,l-1}) + D_{jkl}^{(+,-)}(a)(v_j\otimes f_{k,l}),
\end{aligned}
\tag{37}
$$

then the above four functions are given as follows:

$$
\begin{aligned}
A_{jkl}^{(+,-)}(a) &= \sqrt{c_1c_2}\frac{a_1a_2}{\triangle}(d-j)l\, c_{jkl}(a), \\
B_{jkl}^{(+,-)}(a) &= \sqrt{c_1c_2}\frac{a_1a_2}{\triangle}j(l-s)\, c_{jkl}(a), \\
C_{jkl}^{(+,-)}(a) &= -\frac{1}{2}l\Big[\partial_1 - 4\pi c_1a_1^2 + (b_1-k+l+2r-2s-2) \\
&\qquad +(1+k-l-r+s)\frac{2c_1a_1^2}{\triangle}\Big]c_{jkl}(a), \\
D_{jkl}^{(+,-)}(a) &= \frac{1}{2}(l-s)\Big[\partial_2 - 4\pi c_2a_2^2 + (b_2+k-l-2) \\
&\qquad -(1-k+l)\frac{2c_2a_2^2}{\triangle}\Big]c_{jkl}(a).
\end{aligned}
$$

Here, we put $\partial_i = a_i\frac{\partial}{\partial a_i}$ *: the Euler operator* $(i=1,2)$ *and* $\triangle = (c_1a_1^2 - c_2a_2^2)$.

PROOF. We first note that:

$$\tau \otimes \mathrm{Ad}^{+} = [r,s;u] \otimes [1,1;2] \overset{P^{(-,+)}}{\longrightarrow} [r-1,s+1;u+2],$$

and set $\tau' = [r-1,s+1;u+2]$.

$$\begin{aligned}
&P^{(+,-)} \circ \mathrm{Rad}(\nabla^{+}) c_{jkl}(a)(v_j \otimes f_{kl}) \\
&= \Big[\frac{1}{2}\partial_1 - 2\pi c_1 a_1^2 + \frac{1}{2}\tau'(-\sqrt{-1}H_{13}) \\
&\qquad -3 + \frac{2c_1a_1^2}{\triangle}\Big](c_{jkl}(a)v_j \otimes [-l f_{k+1,l-1}]) \\
&\quad + \Big[\frac{1}{2}\frac{a_1a_2}{\triangle}\chi(X_5 + \sqrt{-1}X_6) + \tau'(e_+^1) \\
&\qquad - \frac{c_1a_1^2}{\triangle}\tau'(e_+^1 + e_+^2)\Big](c_{jkl}(a)v_j \otimes [-l f_{k,l-1}]) \\
&\quad + \Big[\frac{1}{2}\frac{a_1a_2}{\triangle}\chi(X_5 - \sqrt{-1}X_6) - \tau'(e_-^2) \\
&\qquad + \frac{c_1a_1^2}{\triangle}\tau'(e_-^1 + e_-^2)\Big](c_{jkl}(a)v_j \otimes [(l-s) f_{k+1,l}]) \\
&\quad + \Big[\frac{1}{2}\partial_2 - 2\pi c_2 a_2^2 + \frac{1}{2}\tau'(-\sqrt{-1}H_{24}) \\
&\qquad -1 - \frac{2c_1a_1^2}{\triangle}\Big](c_{jkl}(a)v_j \otimes [(l-s) f_{k,l}])
\end{aligned}$$

Here, we use the fact (from Lemma 4.4):

$$\begin{aligned}
P^{(+,-)}(f_{kl} \otimes X_{13}) &= -l f_{k+1,l-1}, & P^{(+,-)}(f_{kl} \otimes X_{23}) &= -l f_{k,l-1}, \\
P^{(+,-)}(f_{kl} \otimes X_{14}) &= (l-s) f_{k+1,l}, & P^{(+,-)}(f_{kl} \otimes X_{24}) &= (l-s) f_{k,l}.
\end{aligned}$$

Next we use the formulas from Lemmas 4.1 and 5.1:

$$\begin{aligned}
&\tau'(-\sqrt{-1}H_{13})(f_{k+1,l-1}) = (b_1 + k - l + 2) f_{k+1,l-1}, \\
&\tau'(e_+^1)(f_{k,l-1}) = (r+1-k) f_{k+1,l-1}, \quad \tau'(e_+^2)(f_{k,l-1}) = (s-l) f_{k,l}, \\
&\tau'(e_-^2)(f_{k+1,l}) = l f_{k+1,l-1}, \qquad \tau'(e_-^1)(f_{k+1,l}) = (k+1) f_{k,l}, \\
&\tau'(-\sqrt{-1}H_{24})(f_{k,l}) = (b_2 - k + l + 2) f_{k,l}.
\end{aligned}$$

and

$$\begin{aligned}
\frac{1}{2}\frac{a_1a_2}{\triangle}\chi(X_5 + \sqrt{-1}X_6)v_j &= -\sqrt{c_1c_2}\frac{a_1a_2}{\triangle}(d-j)v_{j+1}, \\
\frac{1}{2}\frac{a_1a_2}{\triangle}\chi(X_5 - \sqrt{-1}X_6)v_j &= \sqrt{c_1c_2}\frac{a_1a_2}{\triangle} j v_{j-1}.
\end{aligned}$$

Then we have:

$$
\begin{aligned}
&P^{(+,-)} \circ \mathrm{Rad}(\nabla^+) c_{jkl}(a)(v_j \otimes f_{kl}) \\
&= -l \left[\frac{1}{2}\partial_1 - 2\pi c_1 a_1^2 + \frac{1}{2}(b_1 + k - l - 4) \right. \\
&\qquad\qquad \left. + \frac{2c_1 a_1^2}{\triangle} \right] (c_{jkl}(a) v_j \otimes f_{k+1,l-1}) \\
&\quad -l \left[-\sqrt{c_1 c_2}\frac{a_1 a_2}{\triangle}\chi(X_+) + \tau'(e_+^1) \right. \\
&\qquad\qquad \left. - \frac{c_1 a_1^2}{\triangle}\tau'(e_+^1 + e_+^2) \right] (c_{jkl}(a) v_j \otimes f_{k,l-1}) \\
&\quad +(l-s) \left[\sqrt{c_1 c_2}\frac{a_1 a_2}{\triangle}\chi(X_-) - \tau'(e_-^2) \right. \\
&\qquad\qquad \left. + \frac{c_1 a_1^2}{\triangle}\tau'(e_-^1 + e_-^2) \right] (c_{jkl}(a) v_j \otimes f_{k+1,l}) \\
&\quad +(l-s) \left[\frac{1}{2}\partial_2 - 2\pi c_2 a_2^2 + \frac{1}{2}(b_2 - k + l) \right. \\
&\qquad\qquad \left. - \frac{2c_1 a_1^2}{\triangle} \right] (c_{jkl}(a) v_j \otimes f_{k,l}) \\
&= -l \left[\frac{1}{2}\partial_1 - 2\pi c_1 a_1^2 + \frac{1}{2}(b_1 + k - l - 4) \right. \\
&\qquad\qquad \left. + \frac{2c_1 a_1^2}{\triangle} \right] c_{jkl}(a)(v_j \otimes f_{k+1,l-1}) \\
&\quad + l\sqrt{c_1 c_2}\frac{a_1 a_2}{\triangle}(d-j)\, c_{jkl}(a)(v_{j+1} \otimes f_{k,l-1}) \\
&\quad -l(1 - \frac{c_1 a_1^2}{\triangle})(r+1-k)\, c_{jkl}(a)(v_j \otimes f_{k+1,l-1}) \\
&\quad + l\frac{c_1 a_1^2}{\triangle}(s-l)\, c_{jkl}(a)(v_j \otimes f_{k,l}) \\
&\quad +(l-s)\sqrt{c_1 c_2}\frac{a_1 a_2}{\triangle} j\, c_{jkl}(a)(v_{j-1} \otimes f_{k+1,l}) \\
&\quad +(l-s)(-1 + \frac{c_1 a_1^2}{\triangle}) l\, c_{jkl}(a)(v_j \otimes f_{k+1,l-1}) \\
&\quad +(l-s)\frac{c_1 a_1^2}{\triangle}(k+1)\, c_{jkl}(a)(v_j \otimes f_{k,l}) \\
&\quad +(l-s) \left[\frac{1}{2}\partial_2 - 2\pi c_2 a_2^2 + \frac{1}{2}(b_2 - k + l) - \frac{2c_1 a_1^2}{\triangle} \right] c_{jkl}(a)(v_j \otimes f_{k,l})
\end{aligned}
$$

Therefore we have :

$$
\begin{aligned}
&P^{(+,-)} \circ \mathrm{Rad}(\nabla^+) c_{jkl}(a)(v_j \otimes f_{kl}) \\
&= \sqrt{c_1 c_2}\frac{a_1 a_2}{\triangle}(d-j)l\, c_{jkl}(a)(v_{j+1} \otimes f_{k,l-1}) \\
&\quad + \sqrt{c_1 c_2}\frac{a_1 a_2}{\triangle} j(l-s)\, c_{jkl}(a)(v_{j-1} \otimes f_{k+1,l}) \\
&\quad - l\left[\frac{1}{2}\partial_1 - 2\pi c_1 a_1^2 + \frac{1}{2}(b_1 + k - l - 4) + \frac{2c_1 a_1^2}{\triangle} + (r+1-k)\right. \\
&\qquad \left. -(r+1-k)\frac{c_1 a_1^2}{\triangle} + (l-s) - (l-s)\frac{c_1 a_1^2}{\triangle}\right] c_{jkl}(a)(v_j \otimes f_{k+1,l-1}) \\
&\quad + (l-s)\left[\frac{1}{2}\partial_2 - 2\pi c_2 a_2^2 + \frac{1}{2}(b_2 - k + l) - \frac{2c_1 a_1^2}{\triangle}\right. \\
&\qquad \left. - l\frac{c_1 a_1^2}{\triangle} + (k+1)\frac{c_1 a_1^2}{\triangle}\right] c_{jkl}(a)(v_j \otimes f_{k,l}).
\end{aligned}
$$

So we get the desired formula. □

7.4. $\bar{P}^{(-,-)} \circ \mathrm{Rad}(\nabla^-) c_{jkl}(a)(v_j \otimes f_{kl})$.

LEMMA 7.4. *If we write*

$$
\begin{aligned}
&\bar{P}^{(-,-)} \circ \mathrm{Rad}(\nabla^-) c_{jkl}(a)(v_j \otimes f_{kl}) \\
&= \bar{A}_{jkl}^{(-,-)}(a)(v_{j+1} \otimes f_{k-1,l-1}) + \bar{B}_{jkl}^{(-,-)}(a)(v_{j-1} \otimes f_{k,l}) \\
&\quad + \bar{C}_{jkl}^{(-,-)}(a)(v_j \otimes f_{k-1,l}) + \bar{D}_{jkl}^{(-,-)}(a)(v_j \otimes f_{k,l-1}),
\end{aligned}
\tag{38}
$$

then the above four functions are given as follows:

$$
\begin{aligned}
\bar{A}_{jkl}^{(-,-)}(a) &= -\sqrt{c_1 c_2}\frac{a_1 a_2}{\triangle}(d-j)kl\, c_{jkl}(a), \\
\bar{B}_{jkl}^{(-,-)}(a) &= \sqrt{c_1 c_2}\frac{a_1 a_2}{\triangle} j(k-r)(s-l)\, c_{jkl}(a), \\
\bar{C}_{jkl}^{(-,-)}(a) &= \frac{1}{2}k(l-s)\big[\partial_1 + 4\pi c_1 a_1^2 - (b_1 - k + l + 2r + 4) \\
&\qquad + (2 + r - k + l)\frac{2c_1 a_1^2}{\triangle}\big] c_{jkl}(a), \\
\bar{D}_{jkl}^{(-,-)}(a) &= \frac{1}{2}(r-k)l\big[\partial_2 + 4\pi c_2 a_2^2 - (b_2 + k - l + 2s + 4) \\
&\qquad - (2 + k + s - l)\frac{2c_2 a_2^2}{\triangle}\big] c_{jkl}(a).
\end{aligned}
$$

Here, we put $\partial_i = a_i \frac{\partial}{\partial a_i}$ *: the Euler operator* $(i = 1, 2)$ *and* $\triangle = (c_1 a_1^2 - c_2 a_2^2)$.

PROOF. We first note that:

$$\tau \otimes \mathrm{Ad}^- = [r,s;u] \otimes [1,1;-2] \overset{\bar{P}^{(-,-)}}{\longrightarrow} [r-1,s-1;u-2],$$

and set $\tau' = [r-1,s-1;u-2]$.

$$\begin{aligned}
&\bar{P}^{(-,-)} \circ \mathrm{Rad}(\nabla^-) c_{jkl}(a)(v_j \otimes f_{kl}) \\
&= \Big[\frac{1}{2}\partial_1 + 2\pi c_1 a_1^2 - \frac{1}{2}\tau'(-\sqrt{-1}H_{13}) \\
&\qquad -3 + \frac{2c_1a_1^2}{\triangle}\Big] (c_{jkl}(a)v_j \otimes [k(l-s)f_{k-1,l}]) \\
&\quad + \Big[\frac{1}{2}\frac{a_1a_2}{\triangle}\chi(X_5+\sqrt{-1}X_6) + \tau'(e_+^2) \\
&\qquad - \frac{c_1a_1^2}{\triangle}\tau'(e_+^1+e_+^2)\Big] (c_{jkl}(a)v_j \otimes [klf_{k-1,l-1}]) \\
&\quad + \Big[\frac{1}{2}\frac{a_1a_2}{\triangle}\chi(X_5-\sqrt{-1}X_6) - \tau'(e_-^1) \\
&\qquad + \frac{c_1a_1^2}{\triangle}\tau'(e_-^1+e_-^2)\Big] (c_{jkl}(a)v_j \otimes [(k-r)(s-l)f_{k,l}]) \\
&\quad + \Big[\frac{1}{2}\partial_2 + 2\pi c_2 a_2^2 - \frac{1}{2}\tau'(-\sqrt{-1}H_{24}) \\
&\qquad -1 - \frac{2c_1a_1^2}{\triangle}\Big] (c_{jkl}(a)v_j \otimes [(r-k)lf_{k,l-1}])
\end{aligned}$$

Here, we use the fact (from Lemma 4.4):

$$\begin{aligned}
\bar{P}^{(-,-)}(f_{kl} \otimes X_{31}) &= k(l-s)f_{k-1,l}, \\
\bar{P}^{(-,-)}(f_{kl} \otimes X_{41}) &= klf_{k-1,l-1}, \\
\bar{P}^{(-,-)}(f_{kl} \otimes X_{32}) &= (k-r)(s-l)f_{k,l}, \\
\bar{P}^{(-,-)}(f_{kl} \otimes X_{42}) &= (r-k)lf_{k,l-1}.
\end{aligned}$$

Next we use the formulas from Lemmas 4.1 and 5.1:

$$\begin{aligned}
&\tau'(-\sqrt{-1}H_{13})(f_{k-1,l}) = (b_1+k-l-2)f_{k-1,l}, \\
&\tau'(e_+^2)(f_{k-1,l-1}) = (s-l)f_{k-1,l}, \quad \tau'(e_+^1)(f_{k-1,l-1}) = (r-k)f_{k,l-1}, \\
&\tau'(e_-^1)(f_{k,l}) = kf_{k-1,l}, \qquad \tau'(e_-^2)(f_{k,l}) = lf_{k,l-1}, \\
&\tau'(-\sqrt{-1}H_{24})(f_{k,l-1}) = (b_2-k+l-2)f_{k,l-1}.
\end{aligned}$$

and

$$\frac{1}{2}\frac{a_1a_2}{\triangle}\chi(X_5+\sqrt{-1}X_6)v_j = -\sqrt{c_1c_2}\frac{a_1a_2}{\triangle}(d-j)v_{j+1},$$

$$\frac{1}{2}\frac{a_1a_2}{\triangle}\chi(X_5-\sqrt{-1}X_6)v_j=\sqrt{c_1c_2}\frac{a_1a_2}{\triangle}jv_{j-1}.$$

Then we have:

$$\begin{aligned}
&\bar{P}^{(-,-)}\circ \mathrm{Rad}(\nabla^-)c_{jkl}(a)(v_j\otimes f_{kl})\\
&=k(l-s)\left[\frac{1}{2}\partial_1+2\pi c_1a_1^2-\frac{1}{2}(b_1+k-l+4)\right.\\
&\qquad\left.+\frac{2c_1a_1^2}{\triangle}\right](c_{jkl}(a)v_j\otimes f_{k-1,l})\\
&\quad+kl\left[-\sqrt{c_1c_2}\frac{a_1a_2}{\triangle}\chi(X_+)+\tau'(e_+^2)\right.\\
&\qquad\left.-\frac{c_1a_1^2}{\triangle}\tau'(e_+^1+e_+^2)\right](c_{jkl}(a)v_j\otimes f_{k-1,l-1})\\
&\quad+(k-r)(s-l)\left[\sqrt{c_1c_2}\frac{a_1a_2}{\triangle}\chi(X_-)-\tau'(e_-^1)\right.\\
&\qquad\left.+\frac{c_1a_1^2}{\triangle}\tau'(e_-^1+e_-^2)\right](c_{jkl}(a)v_j\otimes f_{k,l})\\
&\quad+(r-k)l\left[\frac{1}{2}\partial_2+2\pi c_2a_2^2-\frac{1}{2}(b_2-k+l)\right.\\
&\qquad\left.-\frac{2c_1a_1^2}{\triangle}\right](c_{jkl}(a)v_j\otimes f_{k,l-1})\\
&=k(l-s)\left[\frac{1}{2}\partial_1+2\pi c_1a_1^2-\frac{1}{2}(b_1+k-l+4)\right.\\
&\qquad\left.+\frac{2c_1a_1^2}{\triangle}\right]c_{jkl}(a)(v_j\otimes f_{k-1,l})\\
&\quad-kl\sqrt{c_1c_2}\frac{a_1a_2}{\triangle}(d-j)\,c_{jkl}(a)(v_{j+1}\otimes f_{k-1,l})\\
&\quad+kl(1-\frac{c_1a_1^2}{\triangle})(s-l)\,c_{jkl}(a)(v_j\otimes f_{k-1,l})\\
&\quad-kl\frac{c_1a_1^2}{\triangle}(r-k)\,c_{jkl}(a)(v_j\otimes f_{k,l-1})\\
&\quad+(k-r)(s-l)\sqrt{c_1c_2}\frac{a_1a_2}{\triangle}j\,c_{jkl}(a)(v_{j-1}\otimes f_{k,l+1})\\
&\quad+(k-r)(s-l)(-1+\frac{c_1a_1^2}{\triangle})k\,c_{jkl}(a)(v_j\otimes f_{k-1,l})\\
&\quad+(k-r)(s-l)\frac{c_1a_1^2}{\triangle}l\,c_{jkl}(a)(v_j\otimes f_{k,l-1})
\end{aligned}$$

$$+(r-k)l\left[\frac{1}{2}\partial_2+2\pi c_2a_2^2-\frac{1}{2}(b_2-k+l)\right.$$
$$\left.-\frac{2c_1a_1^2}{\triangle}\right]c_{jkl}(a)(v_j\otimes f_{k,l-1})$$

Therefore we have :

$$\bar{P}^{(-,-)}\circ\mathrm{Rad}(\nabla^-)c_{jkl}(a)(v_j\otimes f_{kl})$$
$$=-\sqrt{c_1c_2}\frac{a_1a_2}{\triangle}(d-j)kl\,c_{jkl}(a)(v_{j+1}\otimes f_{k-1,l-1})$$
$$+\sqrt{c_1c_2}\frac{a_1a_2}{\triangle}j(k-r)(s-l)\,c_{jkl}(a)(v_{j-1}\otimes f_{k,l})$$
$$+k(l-s)\left[\frac{1}{2}\partial_1+2\pi c_1a_1^2-\frac{1}{2}(b_1+k-l+4)+\frac{2c_1a_1^2}{\triangle}-l\right.$$
$$\left.+l\frac{c_1a_1^2}{\triangle}+(k-r)-(k-r)\frac{c_1a_1^2}{\triangle}\right]c_{jkl}(a)(v_j\otimes f_{k-1,l})$$
$$+(r-k)l\left[\frac{1}{2}\partial_2+2\pi c_2a_2^2-\frac{1}{2}(b_2-k+l)-\frac{2c_1a_1^2}{\triangle}\right.$$
$$\left.-k\frac{c_1a_1^2}{\triangle}-(s-l)\frac{c_1a_1^2}{\triangle}\right]c_{jkl}(a)(v_j\otimes f_{k,l-1}).$$

So we get the desired formula. □

7.5. $\bar{P}^{(-,+)}\circ\mathrm{Rad}(\nabla^-)c_{jkl}(a)(v_j\otimes f_{kl})$.

LEMMA 7.5. *If we write*

$$\bar{P}^{(-,+)}\circ\mathrm{Rad}(\nabla^-)c_{jkl}(a)(v_j\otimes f_{kl})$$
$$=\bar{A}_{jkl}^{(-,+)}(a)(v_{j+1}\otimes f_{k-1,l})+\bar{B}_{jkl}^{(-,+)}(a)(v_{j-1}\otimes f_{k,l+1})$$
$$+\bar{C}_{jkl}^{(-,+)}(a)(v_j\otimes f_{k-1,l+1})+\bar{D}_{jkl}^{(-,+)}(a)(v_j\otimes f_{k,l}),\tag{39}$$

then the above four functions are given as follows:

$$\bar{A}_{jkl}^{(-,+)}(a)=\sqrt{c_1c_2}\frac{a_1a_2}{\triangle}(d-j)k\,c_{jkl}(a),$$
$$\bar{B}_{jkl}^{(-,+)}(a)=\sqrt{c_1c_2}\frac{a_1a_2}{\triangle}j(k-r)\,c_{jkl}(a),$$
$$\bar{C}_{jkl}^{(-,+)}(a)=-\frac{1}{2}k\Big[\partial_1+4\pi c_1a_1^2-(b_1-k+l+2r-2s+2)$$
$$+(1+r-k-s+l)\frac{2c_1a_1^2}{\triangle}\Big]c_{jkl}(a),$$
$$\bar{D}_{jkl}^{(-,+)}(a)=\frac{1}{2}(k-r)\Big[\partial_2+4\pi c_2a_2^2-(b_2+k-l+2)$$

$$-(1+k-l)\frac{2c_2a_2^2}{\triangle}\Big]c_{jkl}(a).$$

Here, we put $\partial_i = a_i\frac{\partial}{\partial a_i}$ *: the Euler operator* $(i=1,2)$ *and* $\triangle = (c_1a_1^2 - c_2a_2^2)$.

PROOF. We first note that:

$$\tau\otimes \mathrm{Ad}^- = [r,s;u]\otimes[1,1;-2] \overset{\bar{P}^{(-,+)}}{\longrightarrow} [r-1,s+1;u-2],$$

and set $\tau' = [r-1,s+1;u-2]$.

$$\begin{aligned}
&\bar{P}^{(-,+)}\circ \mathrm{Rad}(\nabla^-)c_{jkl}(a)(v_j\otimes f_{kl})\\
&=\Big[\frac{1}{2}\partial_1 + 2\pi c_1a_1^2 - \frac{1}{2}\tau'(-\sqrt{-1}H_{13})\\
&\qquad -3+\frac{2c_1a_1^2}{\triangle}\Big]\,(c_{jkl}(a)v_j\otimes[-kf_{k-1,l+1}])\\
&\quad+\Big[\frac{1}{2}\frac{a_1a_2}{\triangle}\chi(X_5+\sqrt{-1}X_6)+\tau'(e_+^2)\\
&\qquad -\frac{c_1a_1^2}{\triangle}\tau'(e_+^1+e_+^2)\Big]\,(c_{jkl}(a)v_j\otimes[-kf_{k-1,l}])\\
&\quad+\Big[\frac{1}{2}\frac{a_1a_2}{\triangle}\chi(X_5-\sqrt{-1}X_6)-\tau'(e_-^1)\\
&\qquad +\frac{c_1a_1^2}{\triangle}\tau'(e_-^1+e_-^2)\Big]\,(c_{jkl}(a)v_j\otimes[(k-r)f_{k,l+1}])\\
&\quad+\Big[\frac{1}{2}\partial_2 + 2\pi c_2a_2^2 - \frac{1}{2}\tau'(-\sqrt{-1}H_{24})\\
&\qquad -1-\frac{2c_1a_1^2}{\triangle}\Big]\,(c_{jkl}(a)v_j\otimes[-(r-k)f_{k,l}])
\end{aligned}$$

Here, we use the fact (from Lemma 4.4):

$$\bar{P}^{(-,+)}(f_{kl}\otimes X_{31}) = -kf_{k-1,l+1}, \qquad \bar{P}^{(-,+)}(f_{kl}\otimes X_{41}) = -kf_{k-1,l},$$
$$\bar{P}^{(-,+)}(f_{kl}\otimes X_{32}) = (k-r)f_{k,l+1}, \quad \bar{P}^{(-,+)}(f_{kl}\otimes X_{42}) = (k-r)f_{k,l}.$$

Next we use the formulas from Lemmas 4.1 and 5.1:

$$\tau'(-\sqrt{-1}H_{13})(f_{k-1,l+1}) = (b_1+k-l-2)f_{k-1,l+1},$$
$$\tau'(e_+^2)(f_{k-1,l}) = (1+s-l)f_{k-1,l+1}, \quad \tau'(e_+^1)(f_{k-1,l}) = (r-k)f_{k,l},$$
$$\tau'(e_-^1)(f_{k,l+1}) = kf_{k-1,l+1}, \qquad \tau'(e_-^2)(f_{k,l+1}) = (l+1)f_{k,l},$$
$$\tau'(-\sqrt{-1}H_{24})(f_{k,l}) = (b_2-k+l-2)f_{k,l}.$$

and

$$\frac{1}{2}\frac{a_1a_2}{\triangle}\chi(X_5+\sqrt{-1}X_6)v_j = -\sqrt{c_1c_2}\frac{a_1a_2}{\triangle}(d-j)v_{j+1},$$
$$\frac{1}{2}\frac{a_1a_2}{\triangle}\chi(X_5-\sqrt{-1}X_6)v_j = \sqrt{c_1c_2}\frac{a_1a_2}{\triangle}jv_{j-1}.$$

Then we have:

$$\begin{aligned}
&\bar{P}^{(-,+)}\circ \mathrm{Rad}(\nabla^-)c_{jkl}(a)(v_j\otimes f_{kl})\\
&= -k\left[\frac{1}{2}\partial_1 + 2\pi c_1a_1^2 - \frac{1}{2}(b_1+k-l+4)\right.\\
&\qquad\qquad \left.+\frac{2c_1a_1^2}{\triangle}\right](c_{jkl}(a)v_j\otimes f_{k-1,l+1})\\
&\quad -k\left[-\sqrt{c_1c_2}\frac{a_1a_2}{\triangle}\chi(X_+)+\tau'(e_+^2)\right.\\
&\qquad\qquad \left.-\frac{c_1a_1^2}{\triangle}\tau'(e_+^1+e_+^2)\right](c_{jkl}(a)v_j\otimes f_{k-1,l})\\
&\quad +(k-r)\left[\sqrt{c_1c_2}\frac{a_1a_2}{\triangle}\chi(X_-)-\tau'(e_-^1)\right.\\
&\qquad\qquad \left.+\frac{c_1a_1^2}{\triangle}\tau'(e_-^1+e_-^2)\right](c_{jkl}(a)v_j\otimes f_{k,l+1})\\
&\quad +(k-r)\left[\frac{1}{2}\partial_2 + 2\pi c_2a_2^2 - \frac{1}{2}(b_2-k+l)\right.\\
&\qquad\qquad \left.-\frac{2c_1a_1^2}{\triangle}\right](c_{jkl}(a)v_j\otimes f_{k,l})\\
&= -k\left[\frac{1}{2}\partial_1 + 2\pi c_1a_1^2 - \frac{1}{2}(b_1+k-l+4)\right.\\
&\qquad\qquad \left.+\frac{2c_1a_1^2}{\triangle}\right] c_{jkl}(a)(v_j\otimes f_{k-1,l+1})\\
&\quad +k\sqrt{c_1c_2}\frac{a_1a_2}{\triangle}(d-j)c_{jkl}(a)(v_{j+1}\otimes f_{k-1,l})\\
&\qquad -k(1-\frac{c_1a_1^2}{\triangle})(1+s-l)\, c_{jkl}(a)(v_j\otimes f_{k-1,l+1})\\
&\qquad +k\frac{c_1a_1^2}{\triangle}(r-k)c_{jkl}(a)(v_j\otimes f_{k,l})\\
&\quad +(k-r)\sqrt{c_1c_2}\frac{a_1a_2}{\triangle}j\, c_{jkl}(a)(v_{j-1}\otimes f_{k,l+1})
\end{aligned}$$

$$+(k-r)(-1+\frac{c_1a_1^2}{\triangle})k\, c_{jkl}(a)(v_j\otimes f_{k-1,l+1})$$

$$+(k-r)\frac{c_1a_1^2}{\triangle}(l+1)\, c_{jkl}(a)(v_j\otimes f_{k,l})$$

$$+(k-r)\left[\frac{1}{2}\partial_2+2\pi c_2a_2^2-\frac{1}{2}(b_2-k+l)\right.$$
$$\left.-\frac{2c_1a_1^2}{\triangle}\right]\, c_{jkl}(a)(v_j\otimes f_{k,l})$$

Therefore we have :

$$\bar{P}^{(-,+)}\circ \mathrm{Rad}(\nabla^-)c_{jkl}(a)(v_j\otimes f_{kl})$$
$$=\sqrt{c_1c_2}\frac{a_1a_2}{\triangle}(d-j)k\, c_{jkl}(a)(v_{j+1}\otimes f_{k-1,l})$$
$$+\sqrt{c_1c_2}\frac{a_1a_2}{\triangle}j(k-r)\, c_{jkl}(a)(v_{j-1}\otimes f_{k,l+1})$$
$$-k\left[\frac{1}{2}\partial_1+2\pi c_1a_1^2-\frac{1}{2}(b_1+k-l+4)+\frac{2c_1a_1^2}{\triangle}+(1+s-l)\right.$$
$$\left.-(1+s-l)\frac{c_1a_1^2}{\triangle}+(k-r)-(k-r)\frac{c_1a_1^2}{\triangle}\right]\, c_{jkl}(a)(v_j\otimes f_{k-1,l+1})$$
$$+(k-r)\left[\frac{1}{2}\partial_2+2\pi c_2a_2^2-\frac{1}{2}(b_2-k+l)-\frac{2c_1a_1^2}{\triangle}\right.$$
$$\left.-k\frac{c_1a_1^2}{\triangle}+(l+1)\frac{c_1a_1^2}{\triangle}\right]\, c_{jkl}(a)(v_j\otimes f_{k,l}).$$

So we get the desired formula. □

7.6. $\bar{P}^{(+,-)}\circ \mathrm{Rad}(\nabla^-)c_{jkl}(a)(v_j\otimes f_{kl})$.

LEMMA 7.6. *If we write*

$$\bar{P}^{(+,-)}\circ \mathrm{Rad}(\nabla^-)c_{jkl}(a)(v_j\otimes f_{kl})$$
$$=\bar{A}^{(+,-)}_{jkl}(a)(v_{j+1}\otimes f_{k,l-1})+\bar{B}^{(+,-)}_{jkl}(a)(v_{j-1}\otimes f_{k+1,l})$$
$$+\bar{C}^{(+,-)}_{jkl}(a)(v_j\otimes f_{k,l})+\bar{D}^{(+,-)}_{jkl}(a)(v_j\otimes f_{k+1,l-1}), \qquad (40)$$

then the above four functions are given as follows:

$$\bar{A}^{(+,-)}_{jkl}(a)=\sqrt{c_1c_2}\frac{a_1a_2}{\triangle}(d-j)l\, c_{jkl}(a),$$

$$\bar{B}^{(+,-)}_{jkl}(a)=\sqrt{c_1c_2}\frac{a_1a_2}{\triangle}j(l-s)\, c_{jkl}(a),$$

$$\bar{C}^{(+,-)}_{jkl}(a)=\frac{1}{2}(s-l)\Big[\partial_1+4\pi c_1a_1^2-(b_1-k+l+2)$$

$$+(1-k+l)\frac{2c_1a_1^2}{\triangle}\Big]c_{jkl}(a),$$

$$\bar{D}^{(+,-)}_{jkl}(a)=\frac{1}{2}l\Big[\partial_2+4\pi c_2a_2^2-(b_2+k-l-2r+2s+2)$$
$$-(1+k-l-r+s)\frac{2c_2a_2^2}{\triangle}\Big]c_{jkl}(a).$$

Here, we put $\partial_i=a_i\frac{\partial}{\partial a_i}$ *: the Euler operator* $(i=1,2)$ *and* $\triangle=(c_1a_1^2-c_2a_2^2)$.

PROOF. We first note that:

$$\tau\otimes\mathrm{Ad}^-=[r,s;u]\otimes[1,1;-2]\stackrel{\bar{P}^{(+,-)}}{\longrightarrow}[r+1,s-1;u-2],$$

and set $\tau'=[r+1,s-1;u-2]$.

$$\bar{P}^{(+,-)}\circ\mathrm{Rad}(\nabla^-)c_{jkl}(a)(v_j\otimes f_{kl})$$
$$=\left[\frac{1}{2}\partial_1+2\pi c_1a_1^2-\frac{1}{2}\tau'(-\sqrt{-1}H_{13})\right.$$
$$\left.-3+\frac{2c_1a_1^2}{\triangle}\right](c_{jkl}(a)v_j\otimes[(s-l)f_{k,l}])$$
$$+\left[\frac{1}{2}\frac{a_1a_2}{\triangle}\chi(X_5+\sqrt{-1}X_6)+\tau'(e^2_+)\right.$$
$$\left.-\frac{c_1a_1^2}{\triangle}\tau'(e^1_++e^2_+)\right](c_{jkl}(a)v_j\otimes[-lf_{k,l-1}])$$
$$+\left[\frac{1}{2}\frac{a_1a_2}{\triangle}\chi(X_5-\sqrt{-1}X_6)-\tau'(e^1_-)\right.$$
$$\left.+\frac{c_1a_1^2}{\triangle}\tau'(e^1_-+e^2_-)\right](c_{jkl}(a)v_j\otimes[(l-s)f_{k+1,l}])$$
$$+\left[\frac{1}{2}\partial_2+2\pi c_2a_2^2-\frac{1}{2}\tau'(-\sqrt{-1}H_{24})\right.$$
$$\left.-1-\frac{2c_1a_1^2}{\triangle}\right](c_{jkl}(a)v_j\otimes[lf_{k+1,l-1}])$$

Here, we use the fact (from Lemma 4.4):

$$\bar{P}^{(+,-)}(f_{kl}\otimes X_{31})=(s-l)f_{k,l},\qquad \bar{P}^{(+,-)}(f_{kl}\otimes X_{41})=-lf_{k,l-1},$$
$$\bar{P}^{(+,-)}(f_{kl}\otimes X_{32})=(l-s)f_{k+1,l},\qquad \bar{P}^{(+,-)}(f_{kl}\otimes X_{42})=lf_{k+1,l-1}.$$

Next we use the formulas from Lemmas 4.1 and 5.1:

$$\tau'(-\sqrt{-1}H_{13})(f_{k,l})=(b_1+k-l-2)f_{k,l},$$

$$\tau'(e_+^2)(f_{k,l-1}) = (s-l)f_{k,l}, \quad \tau'(e_+^1)(f_{k,l-1}) = (r+1-k)f_{k+1,l-1},$$
$$\tau'(e_-^1)(f_{k+1,l}) = (k+1)f_{k,l}, \qquad \tau'(e_-^2)(f_{k+1,l}) = lf_{k+1,l-1},$$
$$\tau'(-\sqrt{-1}H_{24})(f_{k+1,l-1}) = (b_2-k+l-2)f_{k+1,l-1}.$$

and

$$\frac{1}{2}\frac{a_1a_2}{\triangle}\chi(X_5+\sqrt{-1}X_6)v_j = -\sqrt{c_1c_2}\frac{a_1a_2}{\triangle}(d-j)v_{j+1},$$
$$\frac{1}{2}\frac{a_1a_2}{\triangle}\chi(X_5-\sqrt{-1}X_6)v_j = \sqrt{c_1c_2}\frac{a_1a_2}{\triangle}jv_{j-1}.$$

Then we have:

$$\begin{aligned}
&\bar{P}^{(+,-)} \circ \mathrm{Rad}(\nabla^-)c_{jkl}(a)(v_j\otimes f_{kl})\\
&= (s-l)\left[\frac{1}{2}\partial_1 + 2\pi c_1a_1^2 - \frac{1}{2}(b_1+k-l+4) + \frac{2c_1a_1^2}{\triangle}\right](c_{jkl}(a)v_j\otimes f_{k,l})\\
&\quad -l\Big[-\sqrt{c_1c_2}\frac{a_1a_2}{\triangle}\chi(X_+) + \tau'(e_+^2)\\
&\qquad\qquad -\frac{c_1a_1^2}{\triangle}\tau'(e_+^1+e_+^2)\Big](c_{jkl}(a)v_j\otimes f_{k,l-1})\\
&\quad +(l-s)\Big[\sqrt{c_1c_2}\frac{a_1a_2}{\triangle}\chi(X_-) - \tau'(e_-^1)\\
&\qquad\qquad +\frac{c_1a_1^2}{\triangle}\tau'(e_-^1+e_-^2)\Big](c_{jkl}(a)v_j\otimes f_{k+1,l})\\
&\quad +l\left[\frac{1}{2}\partial_2 + 2\pi c_2a_2^2 - \frac{1}{2}(b_2-k+l) - \frac{2c_1a_1^2}{\triangle}\right](c_{jkl}(a)v_j\otimes f_{k+1,l-1})\\
&= (s-l)\left[\frac{1}{2}\partial_1 + 2\pi c_1a_1^2 - \frac{1}{2}(b_1+k-l+4) + \frac{2c_1a_1^2}{\triangle}\right]c_{jkl}(a)(v_j\otimes f_{k,l})\\
&\quad +l\sqrt{c_1c_2}\frac{a_1a_2}{\triangle}(d-j)c_{jkl}(a)(v_{j+1}\otimes f_{k-1,l})\\
&\qquad -l(1-\frac{c_1a_1^2}{\triangle})(s-l)\,c_{jkl}(a)(v_j\otimes f_{k,l})\\
&\qquad +l\frac{c_1a_1^2}{\triangle}(r+1-k)c_{jkl}(a)(v_j\otimes f_{k+1,l-1})\\
&\quad +(l-s)\sqrt{c_1c_2}\frac{a_1a_2}{\triangle}j\,c_{jkl}(a)(v_{j-1}\otimes f_{k+1,l})\\
&\qquad +(l-s)(-1+\frac{c_1a_1^2}{\triangle})(k+1)\,c_{jkl}(a)(v_j\otimes f_{k,l})
\end{aligned}$$

$$+(l-s)\frac{c_1a_1^2}{\triangle}l\, c_{jkl}(a)(v_j\otimes f_{k+1,l-1})$$

$$+l\left[\frac{1}{2}\partial_2+2\pi c_2a_2^2-\frac{1}{2}(b_2-k+l)-\frac{2c_1a_1^2}{\triangle}\right]c_{jkl}(a)(v_j\otimes f_{k+1,l-1})$$

Therefore we have :

$$\begin{aligned}
&\bar{P}^{(+,-)}\circ \mathrm{Rad}(\nabla^-)c_{jkl}(a)(v_j\otimes f_{kl})\\
&=\sqrt{c_1c_2}\frac{a_1a_2}{\triangle}(d-j)l\, c_{jkl}(a)(v_{j+1}\otimes f_{k,l-1})\\
&\quad+\sqrt{c_1c_2}\frac{a_1a_2}{\triangle}j(l-s)\, c_{jkl}(a)(v_{j-1}\otimes f_{k+1,l})\\
&\quad+(s-l)\left[\frac{1}{2}\partial_1+2\pi c_1a_1^2-\frac{1}{2}(b_1+k-l+4)+\frac{2c_1a_1^2}{\triangle}-l\right.\\
&\qquad\left.-l\frac{c_1a_1^2}{\triangle}+(k+1)-(k+1)\frac{c_1a_1^2}{\triangle}\right]c_{jkl}(a)(v_j\otimes f_{k,l})\\
&\quad+l\left[\frac{1}{2}\partial_2+2\pi c_2a_2^2-\frac{1}{2}(b_2-k+l)-\frac{2c_1a_1^2}{\triangle}\right.\\
&\qquad\qquad\left.+(r+1-k)\frac{c_1a_1^2}{\triangle}+(l-s)\frac{c_1a_1^2}{\triangle}\right]c_{jkl}(a)(v_j\otimes f_{k+1,l-1}).
\end{aligned}$$

So we get the desired formula. □

CHAPTER 3

Generalized Whittaker functions for P_J principal series representations

8. Generalized Whittaker functions for P_J-principal series

8.1. P_J-principal series representations and their corner K-types. The Jacobi parabolic subgroup P_J has the Langlands decomposition as $P_J = M_J A_J N_J$. Here, N_J is the unipotent radical, $\mathrm{Lie}(N_J) = \mathfrak{g}_{2\alpha_1} + \mathfrak{g}_{\alpha_1+\alpha_2} + \mathfrak{g}_{\alpha_1-\alpha_2}$ and

$$M_J \simeq \mathbb{C}^{(1)} \times SU(1,1), \quad A_J = \exp(\mathbb{R}H_1).$$

Let $\chi_m(e^{\sqrt{-1}\theta}) = e^{m\sqrt{-1}\theta}$, $m \in \mathbb{Z}$ be a character of $\mathbb{C}^{(1)}$ and $D_p^{\pm}$ be the discrete series representation of $SU(1,1)$ with Blattner parameter $\pm p$. Then we define a discrete series representation $(\chi_m, D_p^{\pm})$ of M_J. Set $\rho_J = 3\alpha_1$.

For $\nu \in \mathfrak{a}_{J,\mathbb{C}}^*$,

$$\pi_J = \mathrm{Ind}_{P_J}^G((\chi_m, D_p^{\pm}) \otimes e^{\nu+\rho_J} \otimes 1_{N_J})$$

is called P_J principal series representations.

We introduce the notion of corner K-type of π_J which is analogous to the minimal K-type of discrete series representations.

DEFINITION 8.1 (The corner K-type of π_J). Let π_J be as above. The corner K-type τ^* of π_J is characterized by the following property:
(1) $\dim \tau_\mu^*$ is minimum in $\pi_J|_K$
(2) $\tau_\mu^*|_{K_{M_J}}$ has the minimal K_{M_J}-type of $D_p^{\pm}$
(3) If $m \neq 0$, there is a non-compact root δ with respect to $(\mathfrak{g}_\mathbb{C}, \mathfrak{t}_\mathbb{C})$ such that $\tau_{\mu+\delta} \in \hat{K}$ and $[\pi_J|_K : \tau_{\mu+\delta}] = 0$.
Here, K and K_{M_J} are maximal compact subgroups of G and M_J respectively.

Their corner K-types are known:

PROPOSITION 8.2 (Hayata [**Ha2**, pp. 540–541]). *If π_J has τ^* as its corner K-type, then the parameter of $\tau = [r, s; u]$ is given as follows.*

$$D_p^+ \ \textit{and} \quad m > 0 \Longrightarrow [r, s; u] = [0, m; -2p + m],$$

$$
\begin{aligned}
&D_p^+ \text{ and } && m<0 \Longrightarrow [r,s;u]=[-m,0;-2p-m],\\
&D_p^+ \text{ and } && m=0 \Longrightarrow [r,s;u]=[0,0;-2p],\\
&D_p^- \text{ and } && m>0 \Longrightarrow [r,s;u]=[m,0;2p-m],\\
&D_p^- \text{ and } && m<0 \Longrightarrow [r,s;u]=[0,-m;2p+m],\\
&D_p^- \text{ and } && m=0 \Longrightarrow [r,s;u]=[0,0;2p].
\end{aligned}
$$

8.2. Differential equations satisfied by generalized Whittaker functions.

PROPOSITION 8.3 (Hayata [**Ha2**]). *Let* $\pi = \mathrm{Ind}_{P_J}^G((\chi_m, D_p^-)\otimes e^{\nu+\rho_J} \otimes 1_{N_J})$. *Generalized Whittaker functions* $\Phi_{\pi,\tau}$ *with the corner* K*-type* τ^* *satisfy one of the following equations.*

$$
\begin{aligned}
&D_p^+ \text{ and } && m>0 \Longrightarrow P^{(+,-)}\circ\nabla^+\Phi_{\pi,\tau}=0, && (41)\\
&D_p^+ \text{ and } && m<0 \Longrightarrow P^{(-,+)}\circ\nabla^+\Phi_{\pi,\tau}=0, && (42)\\
&D_p^+ \text{ and } && m=0 \Longrightarrow P^{(-,-)}\circ\nabla^+\circ P^{(+,+)}\circ\nabla^+\Phi_{\pi,\tau}=0, && (43)\\
&D_p^- \text{ and } && m>0 \Longrightarrow \bar{P}^{(-,+)}\circ\nabla^-\Phi_{\pi,\tau}=0, && (44)\\
&D_p^- \text{ and } && m<0 \Longrightarrow \bar{P}^{(+,-)}\circ\nabla^-\Phi_{\pi,\tau}=0, && (45)\\
&D_p^- \text{ and } && m=0 \Longrightarrow \bar{P}^{(-,-)}\circ\nabla^-\circ\bar{P}^{(+,+)}\circ\nabla^-\Phi_{\pi,\tau}=0. && (46)
\end{aligned}
$$

PROOF. From Proposition 8.2. □

PROPOSITION 8.4. *Let* π *is as above, then we have*

$$\mathrm{Rad}(\Omega)\Phi_{\pi,\tau} = \chi_\pi(\Omega)\Phi_{\pi,\tau}. \tag{47}$$

The infinitesimal character of $\pi = \mathrm{Ind}_{P_J}^G((\chi_m, D_p^\pm)\otimes e^{\nu+\rho_J}\otimes 1_{N_J})$ *is given by,*

$$\chi_\pi(\Omega) = \nu^2 + (p-1)^2 - 10 + \frac{1}{2}m^2. \tag{48}$$

PROOF. Proposition 3.5 in [**Ha2**]. □

8.3. Explicit differential equations for the case $(D_p^-, m>0)$. We write explicit differential equations of generalized Whittaker functions $\Phi_{\pi,\tau}(a)$ for the case:

⌈ $\pi = \mathrm{Ind}_{P_J}^G((\chi_m, D_p^-)\otimes e^{\nu+\rho_J}\otimes 1_{N_J})$, $(m>0)$ and
with corner K-type $\tau^*, \tau = [m,0;2p-m]$ ⌋.

This is a typical case and the other case is quite similar.

Let $\Phi_{\pi,\tau}(a) = \sum_{j=0}^{d}\sum_{k=0}^{m} c_{jk}(a)(v_j\otimes f_{k0}) \in C_{M_0}^\infty(A, V_\chi\otimes W_\tau)$. Here, $\eta = \chi\otimes\xi$, χ is an irreducible finite-dimensional representation of

$SU(\xi) \simeq SU(2)$ and ξ is a non-trivial unitary character of N_S determined by a positive definite matrix $H_\xi = \begin{pmatrix} c_1 & \\ & c_2 \end{pmatrix}$. We assume that $\dim \chi = d+1$, $\frac{d+m}{2} \in \mathbb{Z}$.

Then the unknown functions $c_{jk}(a)$ satisfy the following difference differential equations.

PROPOSITION 8.5. *Under the above assumptions and notations, the coefficient functions $c_{jk}(a)$ in $\Phi_{\pi,\tau}(a)$ satisfy the following system of partial differential equations. Here, we put $\partial_i = a_i \frac{\partial}{\partial a_i}$: the Euler operator $(i = 1,2)$, $\triangle = (c_1a_1^2 - c_2a_2^2)$ and $c_{j,-1}(a) = c_{j,m+1}(a) = c_{-1,k}(a) = c_{d+1,k}(a) \equiv 0$.*

$$
\begin{aligned}
&2(k+1)\sqrt{c_1c_2}\frac{a_1a_2}{\triangle}(d-j+1)\,c_{j-1,k+1}(a)\\
&\quad -(m-k)\left[\partial_2 + 4\pi c_2a_2^2 - (k+1)\frac{2c_2a_2^2}{\triangle}\right]c_{j,k}(a) = 0, \qquad (49)
\end{aligned}
$$

$$
\begin{aligned}
&2(m-k)\sqrt{c_1c_2}\frac{a_1a_2}{\triangle}j\,c_{j,k}(a) + (k+1)\bigg[\partial_1 + 4\pi c_1a_1^2\\
&\quad -(m-k+p+1) + (m-k)\frac{2c_1a_1^2}{\triangle}\bigg]c_{j-1,k+1}(a) = 0, \qquad (50)
\end{aligned}
$$

and

$$
\begin{aligned}
&\bigg[\partial_1^2 + \partial_2^2 - 2(\partial_1+\partial_2) + 4\frac{c_2a_2^2}{\triangle}\partial_1 - 4\frac{c_1a_1^2}{\triangle}\partial_2 - 16\pi^2(c_1^2a_1^4 + c_2^2a_2^4)\\
&\quad +8\pi c_1a_1^2(p+k-m) + 8\pi c_2a_2^2(p-k)\\
&\quad +\frac{1}{2}(2k-m)^2 - \frac{1}{2}m^2 - \nu^2 - (p-1)^2 + 10\\
&\quad -8c_1c_2\left(\frac{a_1a_2}{\triangle}\right)^2\left\{2k(m-k) + m + \left(\frac{d-m}{2}\right)\left(\frac{d+m}{2}+1\right)\right\}\bigg]c_{j,k}(a)\\
&\quad -4\frac{a_1a_2}{\triangle}\frac{c_1a_1^2 + c_2a_2^2}{\triangle}\sqrt{c_1c_2}(d-j+1)(k+1)\,c_{j-1,k+1}(a)\\
&\quad -4\frac{a_1a_2}{\triangle}\frac{c_1a_1^2 + c_2a_2^2}{\triangle}\sqrt{c_1c_2}(j+1)(m-k+1)\,c_{j+1,k-1}(a)\\
&\quad = 0. \qquad (51)
\end{aligned}
$$

PROOF. We first note that:

$$\tau \otimes \mathrm{Ad}^- = [m,0;2p-m] \otimes [1,1;2] \overset{\bar{P}^{(-,+)}}{\longrightarrow} [m-1,1;2p-m-2],$$

and set $\tau' = [m-1, 1; 2p-m-2]$. Then we use the equation (from Proposition 8.3):

$$\bar{P}^{(-,+)} \circ \mathrm{Rad}(\nabla^-)\Phi_{\pi,\tau}(a) = \sum_{j=0}^{d}\sum_{k=0}^{m} \bar{P}^{(-,+)} \circ \mathrm{Rad}(\nabla^-) c_{jk}(a)(v_j \otimes f_{k0}) = 0.$$

Hence, we calculate the terms $\bar{P}^{(-,+)} \circ \mathrm{Rad}(\nabla^-) c_{jk}(a)(v_j \otimes f_{k0})$ $(0 \le j \le d,\ 0 \le k \le m,\ 2j+2k = d+m)$ by using the Schmid operator.

$$\begin{aligned}
&\bar{P}^{(-,+)} \circ \mathrm{Rad}(\nabla^-) c_{jk}(a)(v_j \otimes f_{k0}) \\
&= \Big[\frac{1}{2}\partial_1 + 2\pi c_1 a_1^2 - \frac{1}{2}\tau'(-\sqrt{-1}H_{13}) \\
&\qquad -3 + \frac{2c_1a_1^2}{\triangle}\Big] (c_{jk}(a)v_j \otimes [-k f_{k-1,1}]) \\
&\quad + \Big[\frac{1}{2}\frac{a_1a_2}{\triangle}\chi(X_5 + \sqrt{-1}X_6) + \tau'(e_+^2) \\
&\qquad - \frac{c_1a_1^2}{\triangle}\tau'(e_+^1 + e_+^2)\Big] (c_{jk}(a)v_j \otimes [-k f_{k-1,0}]) \\
&\quad + \Big[\frac{1}{2}\frac{a_1a_2}{\triangle}\chi(X_5 - \sqrt{-1}X_6) - \tau'(e_-^1) \\
&\qquad + \frac{c_1a_1^2}{\triangle}\tau'(e_-^1 + e_-^2)\Big] (c_{jk}(a)v_j \otimes [(k-m) f_{k,1}]) \\
&\quad + \Big[\frac{1}{2}\partial_2 + 2\pi c_2 a_2^2 - \frac{1}{2}\tau'(-\sqrt{-1}H_{24}) \\
&\qquad -1 - \frac{2c_1a_1^2}{\triangle}\Big] (c_{jk}(a)v_j \otimes [-(m-k) f_{k,0}])
\end{aligned}$$

Here, we use the fact (from Lemma 4.4):

$$\begin{aligned}
&\bar{P}^{(-,+)}(f_{k0} \otimes X_{31}) = -k f_{k-1,1}, &&\bar{P}^{(-,+)}(f_{k0} \otimes X_{41}) = -k f_{k-1,0}, \\
&\bar{P}^{(-,+)}(f_{k0} \otimes X_{32}) = (k-m) f_{k,1}, &&\bar{P}^{(-,+)}(f_{k0} \otimes X_{42}) = -(m-k) f_{k,0}.
\end{aligned}$$

Next we use the formulas from Lemmas 4.1 and 5.1:

$$\begin{aligned}
&\tau'(-\sqrt{-1}H_{13})(f_{k-1,1}) = (k-m+p-2) f_{k-1,1}, \\
&\tau'(e_+^2)(f_{k-1,0}) = f_{k-1,1}, \quad \tau'(e_+^1)(f_{k-1,0}) = (m-k) f_{k,0}, \\
&\tau'(e_-^1)(f_{k,1}) = k f_{k-1,1}, \qquad \tau'(e_-^2)(f_{k,1}) = f_{k,0}, \\
&\tau'(-\sqrt{-1}H_{24})(f_{k,0}) = (p-k-2) f_{k,0}.
\end{aligned}$$

and

$$\frac{1}{2}\frac{a_1a_2}{\triangle}\chi(X_5 + \sqrt{-1}X_6)v_j = -\sqrt{c_1c_2}\frac{a_1a_2}{\triangle}(d-j)v_{j+1},$$

$$\frac{1}{2}\frac{a_1a_2}{\triangle}\chi(X_5-\sqrt{-1}X_6)v_j=\sqrt{c_1c_2}\frac{a_1a_2}{\triangle}jv_{j-1}.$$

Then we have:

$$\begin{aligned}
&\bar{P}^{(-,+)}\circ \mathrm{Rad}(\nabla^-)c_{jk}(a)(v_j\otimes f_{k0})\\
&=-k\left[\frac{1}{2}\partial_1+2\pi c_1a_1^2-\frac{1}{2}(k-m+p-2)\right.\\
&\qquad\qquad\left.-3+\frac{2c_1a_1^2}{\triangle}\right](c_{jk}(a)v_j\otimes f_{k-1,1})\\
&\quad-k\left[-\sqrt{c_1c_2}\frac{a_1a_2}{\triangle}\chi(X_+)+\tau'(e_+^2)\right.\\
&\qquad\qquad\left.-\frac{c_1a_1^2}{\triangle}\tau'(e_+^1+e_+^2)\right](c_{jk}(a)v_j\otimes f_{k-1,0})\\
&\quad+(k-m)\left[\sqrt{c_1c_2}\frac{a_1a_2}{\triangle}\chi(X_-)-\tau'(e_-^1)\right.\\
&\qquad\qquad\left.+\frac{c_1a_1^2}{\triangle}\tau'(e_-^1+e_-^2)\right](c_{jk}(a)v_j\otimes f_{k,1})\\
&\quad+(k-m)\left[\frac{1}{2}\partial_2+2\pi c_2a_2^2-\frac{1}{2}(p-k-2)\right.\\
&\qquad\qquad\left.-1-\frac{2c_1a_1^2}{\triangle}\right](c_{jk}(a)v_j\otimes f_{k,0})\\
&=-k\left[\frac{1}{2}\partial_1+2\pi c_1a_1^2-\frac{1}{2}(k-m+p-2)-3\right.\\
&\qquad\qquad\left.+\frac{2c_1a_1^2}{\triangle}\right]c_{jk}(a)(v_j\otimes f_{k-1,1})\\
&\quad+k\sqrt{c_1c_2}\frac{a_1a_2}{\triangle}(d-j)c_{jk}(a)(v_{j+1}\otimes f_{k-1,0})\\
&\qquad-k(1-\frac{c_1a_1^2}{\triangle})c_{jk}(a)(v_j\otimes f_{k-1,1})\\
&\qquad+k\frac{c_1a_1^2}{\triangle}(m-k)c_{jk}(a)(v_j\otimes f_{k,0})\\
&\quad+(k-m)\sqrt{c_1c_2}\frac{a_1a_2}{\triangle}jc_{jk}(a)(v_{j-1}\otimes f_{k,1})\\
&\qquad+(k-m)(-1+\frac{c_1a_1^2}{\triangle})kc_{jk}(a)(v_j\otimes f_{k-1,1})\\
&\qquad+(k-m)\frac{c_1a_1^2}{\triangle}c_{jk}(a)(v_j\otimes f_{k,0})
\end{aligned}$$

$$+(k-m)\left[\frac{1}{2}\partial_2+2\pi c_2a_2^2-\frac{1}{2}(p-k-2)-1\right.$$
$$\left.-\frac{2c_1a_1^2}{\triangle}\right](c_{jk}(a)v_j\otimes f_{k,0}).$$

Here, we write the last formula as

$$A_{jk}(a)\,(v_{j+1}\otimes f_{k-1,0})+B_{jk}(a)\,(v_{j-1}\otimes f_{k,1})$$
$$+C_{jk}(a)\,(v_j\otimes f_{k-1,1})+D_{jk}(a)\,(v_j\otimes f_{k,0}).$$

Then, four functions are given as follows:

$$A_{jk}(a)=k\sqrt{c_1c_2}\frac{a_1a_2}{\triangle}(d-j)c_{jk}(a),$$
$$B_{jk}(a)=(k-m)\sqrt{c_1c_2}\frac{a_1a_2}{\triangle}jc_{jk}(a),$$
$$C_{jk}(a)=-k\left[\frac{1}{2}\partial_1+2\pi c_1a_1^2-\frac{1}{2}(k-m+p-2)+(k-m-2)\right.$$
$$\left.-(k-m-1)\frac{c_1a_1^2}{\triangle}\right]c_{jk}(a),$$
$$D_{jk}(a)=(k-m)\left[\frac{1}{2}\partial_2+2\pi c_2a_2^2-\frac{1}{2}(p-k-2)-1\right.$$
$$\left.-(k+1)\frac{c_1a_1^2}{\triangle}\right]c_{jk}(a).$$

Therefore,

$$\bar{P}^{(-,+)}\circ\mathrm{Rad}(\nabla^-)\Phi_{\pi,\tau}(a)$$
$$=\sum_{j=0}^{d}\sum_{k=0}^{m}\bar{P}^{(-,+)}\circ\mathrm{Rad}(\nabla)c_{jk}(a)(v_j\otimes f_{k0})=0$$

implies that

$$\sum_{j=0}^{d}\sum_{k=0}^{m}[A_{jk}(a)(v_{j+1}\otimes f_{k-1,0})+D_{jk}(a)(v_j\otimes f_{k,0})]=0$$

and

$$\sum_{j=0}^{d}\sum_{k=0}^{m}[B_{jk}(a)(v_{j-1}\otimes f_{k,1})+C_{jk}(a)(v_j\otimes f_{k-1,1})]=0.$$

So we have

$$A_{j-1,k+1}(a)+D_{j,k}(a)=0\quad\text{and}\quad B_{j,k}(a)+C_{j-1,k+1}(a)=0.\tag{52}$$

They are just (49) and (50).

The other hand, from the equation in Proposition 8.4:

$$\begin{aligned}&\mathrm{Rad}(\Omega)\Phi_{\pi,\tau}(a)-\chi_\pi(\Omega)\Phi_{\pi,\tau}(a)\\&=\sum_{j=0}^{d}\sum_{k=0}^{m}\left[\mathrm{Rad}(\Omega)-\{\nu^2+(p-1)^2-10+\frac{1}{2}m^2\}\right]c_{jk}(a)(v_j\otimes f_{k0})=0.\end{aligned}$$

We calculate the terms $[\mathrm{Rad}(\Omega)-\{\nu^2+(p-1)^2-10+\frac{1}{2}m^2\}]c_{jk}(a)(v_j\otimes f_{k0})$ by using the Casimir operator. Then we get the desired difference-differential equation (51) similar to the case of the Schmid operator. □

8.4. Reduction of the system of partial differential equations. We reduce the system of partial differential equations of the unknown functions $c_{jk}(a)$ in Proposition 8.5 to a simpler holonomic system step by step.

We firstly reduce the system to the following.

PROPOSITION 8.6. *Under the same assumptions and notations in Proposition 8.5, if we write* $c_{jk}(a)=(\sqrt{c_1}a_1)^{m-k+p+2}(\sqrt{c_2}a_2)^{k+p+2}\times e^{-2\pi(c_1a_1^2+c_2a_2^2)}h_{jk}(a)$, *then the functions* $h_{jk}(a)$ *satisfy the following system of partial differential equations.*

$$\begin{aligned}&2(k+1)\frac{c_2a_2^2}{\triangle}(d-j+1)\,h_{j-1,k+1}(a)\\&\qquad-(m-k)\left[\partial_2-(1+k)\frac{2c_2a_2^2}{\triangle}\right]h_{j,k}(a)=0,\end{aligned}\tag{53}$$

$$\begin{aligned}&2(m-k)\frac{c_1a_1^2}{\triangle}j\,h_{j,k}(a)\\&\qquad+(k+1)\left[\partial_1+(m-k)\frac{2c_1a_1^2}{\triangle}\right]h_{j-1,k+1}(a)=0,\end{aligned}\tag{54}$$

and

$$\begin{aligned}&[\partial_1^2+\partial_2^2+2(m+p+1)(\partial_1+\partial_2)-8\pi(c_1a_1^2\partial_1+c_2a_2^2\partial_2)\\&+(k+1)\frac{4c_2a_2^2}{\triangle}\partial_1-(m-k+1)\frac{4c_1a_1^2}{\triangle}\partial_2\\&-8\pi(m+2)(c_1a_1^2+c_2a_2^2)-8\pi(m-2k)(c_1a_1^2-c_2a_2^2)\\&-8c_1c_2(\frac{a_1a_2}{\triangle})^2(\frac{d-m}{2})(\frac{d+m}{2}+1)\\&+(m+p+1)^2-\nu^2]\,h_{j,k}(a)=0.\end{aligned}\tag{55}$$

PROOF. By direct computation. □

Hereafter, we assume that $h_{jk}(a)$ is *analytic* on $A \simeq (\mathbb{R}_{>0})^2$, and study analytic solutions for the above system of differential equations. We require a lemma on analyticity of $h_{jk}(a)$.

LEMMA 8.7. *Under the same assumptions and notations in Proposition 8.6, if analytic function $h_{jk}(a)$ satisfies (53) and (54), then $(c_1a_1^2 - c_2a_2^2)^{-\frac{d-m}{2}} \cdot h_{jk}(a)$ is also analytic on A. In particular $h_{jk}(a)$ is exactly divisible by $(c_1a_1^2 - c_2a_2^2)^{\frac{d-m}{2}}$ when $d > m$.*

PROOF. If $d \leq m$ then $(c_1a_1^2 - c_2a_2^2)^{-\frac{d-m}{2}}$ is a polynomial and this assertion is trivial. We may assume that $d > m$. We introduce new variables $x = 2\pi(c_1a_1^2 - c_2a_2^2)$ and $y = 2\pi(c_1a_1^2 + c_2a_2^2)$. Then $h_{jk}(x, y) = h_{jk}(a_1, a_2)$ is also analytic on the variable x, (note that c_1, c_2 are positive), i.e. there is a integer $N_0 \geq 0$ such that

$$h_{jk}(x, y) = \sum_{n=N_0}^{\infty} H_{jk,n}(y)x^n, \tag{56}$$

for suitable functions $\{H_{jk,n}(y)\}_{n=N_0}^{\infty}$. Next we substitute (56) into (53)–(54). By using the formulas:

$$\partial_1 = (x+y)(\frac{\partial}{\partial x} + \frac{\partial}{\partial y}), \quad \partial_2 = (x-y)(\frac{\partial}{\partial x} - \frac{\partial}{\partial y}),$$

and we have

$$\det\begin{pmatrix} (m-k)(1+k+N_0) & (k+1)(d-j+1) \\ (m-k)j & (k+1)(m-k+N_0) \end{pmatrix} = 0. \tag{57}$$

Since $2(j+k) = d+m$, we have $(N_0 - \frac{d-m}{2})(N_0 + \frac{d+m}{2} + 1) = 0$. $N_0 \geq 0$ implies that $N_0 = \frac{d-m}{2}$. □

We introduce a more tractable function $f_{jk}(a)$ by using the above lemma.

PROPOSITION 8.8. *Under the same assumptions and notations until the above in this subsection, if we write an analytic function $h_{jk}(a) = \triangle^{\frac{d-m}{2}} f_{jk}(a)$ then $f_{jk}(a)$ is also analytic and satisfies the following two partial differential equations.*

$$\Big[\partial_1\partial_2 - \big(\frac{d-m}{2} + k + 1\big)\frac{2c_2a_2^2}{c_1a_1^2 - c_2a_2^2}\partial_1 + \big(\frac{d-m}{2} + m - k + 1\big)\frac{2c_1a_1^2}{c_1a_1^2 - c_2a_2^2}\partial_2\Big] f_{jk}(a) = 0, \tag{58}$$

and

$$\Big[(\partial_1 + \partial_2)^2 + 2(d+p+1)(\partial_1 + \partial_2) - 8\pi(c_1a_1^2\partial_1 + c_2a_2^2\partial_2)$$

$$-8\pi(d+2)(c_1a_1^2+c_2a_2^2)-8\pi(m-2k)(c_1a_1^2-c_2a_2^2)$$
$$+(d+p+1)^2-\nu^2\Big]\, f_{jk}(a)=0. \tag{59}$$

Remark. If we add one more recurrence relation (54) between $f_{j,k}(a)$ and $f_{j-1,k+1}(a)$ to this proposition, then it is equivalent to Proposition 8.5.

PROOF. By using two equations (53) and (54) in Proposition 8.6, we eliminate $h_{j-1,k+1}(a)$. Then we have,

$$\left[\partial_1+(m-k)\frac{2c_1a_1^2}{\triangle}\right]\frac{\triangle}{a_2^2}\left[\partial_2-(1+k)\frac{2c_2a_2^2}{\triangle}\right] h_{jk}(a)$$
$$+4c_1c_2\frac{a_1^2}{\triangle}j(d-j+1)\,h_{jk}(a)=0$$

Thus we get:

$$\Big[\partial_1\partial_2-(1+k)\frac{2c_2a_2^2}{\triangle}\partial_1+(1+m-k)\frac{2c_1a_1^2}{\triangle}\partial_2$$
$$+4c_1c_2(\frac{a_1a_2}{\triangle})^2\{j(d-j+1)-(m-k)(1+k)\}\Big]\,h_{jk}(a)=0. \tag{60}$$

Next we calculate (55)+2×(60) and have:

$$\Big[(\partial_1+\partial_2)^2+2(m+p+1)(\partial_1+\partial_2)-8\pi(c_1a_1^2\partial_1+c_2a_2^2\partial_2)$$
$$-8\pi(m+2)(c_1a_1^2+c_2a_2^2)-8\pi(m-2k)(c_1a_1^2-c_2a_2^2)$$
$$+(m+p+1)^2-\nu^2\Big]\,h_{jk}(a)=0, \tag{61}$$

To get the desired equations (58) and (59), we substitute $h_{jk}(a)=\triangle^{\frac{d-m}{2}}f_{jk}(a)$ in (60) and (61). We use the following formula in this calculation.

Set $\alpha=d-m$, then we have

$$\partial_1(\triangle^{\frac{\alpha}{2}}f(a))=\frac{\alpha}{2}\triangle^{\frac{\alpha}{2}-1}(2c_1a_1^2)f(a)+\triangle^{\frac{\alpha}{2}}(\partial_1 f(a)),$$
$$\partial_2(\triangle^{\frac{\alpha}{2}}f(a))=\frac{\alpha}{2}\triangle^{\frac{\alpha}{2}-1}(-2c_2a_2^2)f(a)+\triangle^{\frac{\alpha}{2}}(\partial_2 f(a)),$$
$$(\partial_1+\partial_2)(\triangle^{\frac{\alpha}{2}}f(a))=\triangle^{\frac{\alpha}{2}}[(\partial_1+\partial_2)+\alpha]f(a),$$

and

$$2\partial_1\partial_2(\triangle^{\frac{\alpha}{2}}f(a))$$
$$=\triangle^{\frac{\alpha}{2}}\left[2\partial_1\partial_2-\alpha\frac{2c_2a_2^2}{\triangle}\partial_1+\alpha\frac{2c_1a_1^2}{\triangle}\partial_2-2\alpha(\alpha-2)c_1c_2(\frac{a_1a_2}{\triangle})^2\right]f(a).$$

Thus we have

$$\Big[\partial_1\partial_2 - \Big(\frac{d-m}{2}+k+1\Big)\frac{2c_2a_2^2}{c_1a_1^2-c_2a_2^2}\partial_1 + \Big(\frac{d-m}{2}+m-k+1\Big)\frac{2c_1a_1^2}{c_1a_1^2-c_2a_2^2}\partial_2 + 4c_1c_2(\frac{a_1a_2}{\triangle})^2\{j(d-j+1)-(m-k)(1+k)\} + 4c_1c_2(\frac{a_1a_2}{\triangle})^2\Big\{-\frac{d-m}{2}(1+k)-\frac{d-m}{2}(1+m-k) - \frac{d-m}{2}\Big(\frac{d-m}{2}-1\Big)\Big\}\Big] f_{jk}(a) = 0,$$

Since $j+k=\frac{d+m}{2}$ implies

$$j(d-j+1)-(m-k)(1+k)-\frac{d-m}{2}(\frac{d+m}{2}+1)=0,$$

The above differential equation becomes (58). The proof of (59) is similar. □

The differential equation (58) in Proposition 8.8 is essentially the *Euler-Darboux* equation which appears in hypergeometric function theory. To see it directly, we introduce new variables $y=(y_1,y_2)=(2\pi c_1a_1^2, 2\pi c_2a_2^2)$.

PROPOSITION 8.9. *After the change of variables, the partial differential equations in Proposition 8.8 is rewritten as follows,*

$$\Big[\frac{\partial^2}{\partial y_1\partial y_2} - (d-j+1)\frac{1}{y_1-y_2}\frac{\partial}{\partial y_1} + (j+1)\frac{1}{y_1-y_2}\frac{\partial}{\partial y_2}\Big] f_{jk}(y) = 0, \tag{62}$$

and

$$\Big[y_1^2\frac{\partial^2}{\partial y_1^2} + y_2^2\frac{\partial^2}{\partial y_2^2} + 2y_1y_2\frac{\partial^2}{\partial y_1\partial y_2} + (d+p+2)(y_1\frac{\partial}{\partial y_1}+y_2\frac{\partial}{\partial y_2}) - 2\Big\{(y_1^2\frac{\partial}{\partial y_1}+y_2^2\frac{\partial}{\partial y_2}) + (\frac{d}{2}+1)(y_1+y_2) + (\frac{m}{2}-k)(y_1-y_2)\Big\} + \frac{1}{4}\{(d+p+1)^2-\nu^2\}\Big] f_{jk}(y) = 0. \tag{63}$$

PROOF. Notc that

$$\partial_1 f = a_1\frac{\partial f}{\partial a_1} = a_1\frac{\partial f}{\partial y_1}\frac{\partial y_1}{\partial a_1} = 4\pi c_1a_1^2\frac{\partial f}{\partial y_1} = 2y_1\frac{\partial f}{\partial y_1},$$

and $\frac{d-m}{2}+k=d-j$, $\frac{d+m}{2}-k=j$. □

We write (62), (63) as

$$P_A\, f_{jk}(y) = 0, \qquad P_B\, f_{jk}(y) = 0, \tag{64}$$

i.e. P_A, P_B are partial differential operators as is defined in Proposition 8.9.

$$P_A = \frac{\partial^2}{\partial y_1 \partial y_2} - (d-j+1)\frac{1}{y_1 - y_2}\frac{\partial}{\partial y_1} + (j+1)\frac{1}{y_1 - y_2}\frac{\partial}{\partial y_2},$$

$$\begin{aligned} P_B = {} & y_1^2\frac{\partial^2}{\partial y_1^2} + y_2^2\frac{\partial^2}{\partial y_2^2} + 2y_1y_2\frac{\partial^2}{\partial y_1\partial y_2} + (d+p+2)(y_1\frac{\partial}{\partial y_1} + y_2\frac{\partial}{\partial y_2}) \\ & -2\left\{(y_1^2\frac{\partial}{\partial y_1} + y_2^2\frac{\partial}{\partial y_2}) + (\frac{d}{2}+1)(y_1+y_2) + (\frac{m}{2}-k)(y_1-y_2)\right\} \\ & +\frac{1}{4}\left\{(d+p+1)^2 - \nu^2\right\}. \end{aligned} \tag{65}$$

LEMMA 8.10 (Debiard etc. [**DG1, DG2**], Iida [**Ii**, Lemma 8.4]). *For the function $f_{jk}(y_1, y_2)$ which belongs to* $\ker P_A$*, if there is a real number r such that $0 < r < \infty$ and $f_{jk}(y_1, y_2)$ is analytic around the point (r, r), then $f_{jk}(y_1, y_2)$ has the following integral expression.*

$$f_{jk}(y) = \int_0^1 F_{jk}(ty_1 + (1-t)y_2)t^j(1-t)^{d-j}\, dt. \tag{66}$$

Here F_{jk} is some analytic function around the point r.

Our function $f_{jk}(y_1, y_2)$ satisfies the assumption in the above lemma. Actually $f_{jk}(y_1, y_2)$ is analytic on $(\mathbb{R}_{>0})^2$. Let us determine the integrand F_{jk}.

LEMMA 8.11. *Let P_B be the differential operator defined in (64). For*

$$f_{jk}(y) = \int_0^1 F_{jk}(ty_1 + (1-t)y_2)t^j(1-t)^{d-j}\, dt$$

Put

$$(P_B\, f_{jk})(y) = \int_0^1 G_{jk}(ty_1 + (1-t)y_2)t^j(1-t)^{d-j}\, dt.$$

Then we have

$$\begin{aligned} G_{jk}(x) = \Big[& x^2\frac{d^2}{dx^2} - \{2x - (d+p+2)\}x\frac{d}{dx} - 2(d+2)x \\ & +\frac{1}{4}\{(d+p+1)^2 - \nu^2\}\Big] F_{jk}(x). \end{aligned} \tag{67}$$

PROOF. By interchanging integration and differentiation. □

Concerning Proposition 8.9, Lemma 8.10 and Lemma 8.11, we have the following proposition.

PROPOSITION 8.12. *Under the same assumptions and notations in Proposition 8.9, The analytic function $f_{jk}(y)$ has the following integral expression*

$$f_{jk}(y) = \int_0^1 F_{jk}(ty_1 + (1-t)y_2)t^j(1-t)^{d-j}\,dt$$

and $F_{jk}(x)$ satisfies the following differential equation

$$\left[x^2\frac{d^2}{dx^2} - \{2x - (d+p+2)\}x\frac{d}{dx} - 2(d+2)x + \frac{1}{4}\{(d+p+1)^2 - \nu^2\}\right] F_{jk}(x) = 0. \tag{68}$$

The next lemma tells us a explicit formula for $f_{jk}(y)$.

LEMMA 8.13. *The differential equation:*

$$\left[x^2\frac{d^2}{dx^2} - \{2x - (d+p+2)\}x\frac{d}{dx} - 2(d+2)x + \frac{1}{4}\{(d+p+1)^2 - \nu^2\}\right] F(x) = 0 \tag{69}$$

has two linearly independent solutions:

$$F_W(x) = e^x x^{-\frac{d+p+2}{2}} W_{\frac{p-d-2}{2},\frac{\nu}{2}}(2x) \tag{70}$$

and

$$F_M(x) = e^x x^{-\frac{d+p+2}{2}} M_{\frac{p-d-2}{2},\frac{\nu}{2}}(2x). \tag{71}$$

Here, $W_{\kappa,\mu}(z)$ and $M_{\kappa,\mu}(z)$ are Whittaker's confluent hypergeometric functions. (See [**WW**] *Chapter 16 for definition.)*

PROOF. Put $F(x) = e^x x^{-\frac{d+p+2}{2}} H(x)$. We transform the differential equation to that of $H(x)$. Then $H(x)$ satisfies:

$$\frac{d^2H(x)}{dx^2} + \left\{-1 + \frac{p-d-2}{x} + \frac{\frac{1}{4} - \frac{\nu^2}{4}}{x^2}\right\} H(x) = 0.$$

If we set $t = 2x$, then

$$\frac{d^2H}{dt^2} + \left\{-\frac{1}{4} + \frac{\frac{p-d-2}{2}}{t} + \frac{\frac{1}{4} - \frac{\nu^2}{4}}{t^2}\right\} H = 0.$$

This is precise Whittaker's differential equation. □

8.5. Main theorem for P_J-principal series. We summarize the results.

THEOREM 8.14. *Let* $\pi = \mathrm{Ind}_{P_J}^G((\chi_m, D_p^-) \otimes e^{\nu+\rho_J} \otimes 1_{N_J})$ *with* $m > 0$*, a positive definite character* $\xi = \mathrm{diag}(c_1, c_2)$ *of* N_S *and* χ *be an irreducible finite-dimensional representation of* $SU(\xi)$ ($\simeq SU(2)$) *with* $\dim \chi = d+1$*. Set* $\eta = \chi \otimes \xi$*.*

If $d < m$ *then*

$$\dim_{\mathbb{C}} \mathrm{Hom}_{(\mathfrak{g},K)}(\pi, C_\eta^\infty(R\backslash G)) = 0,$$

and if $d \geq m$ *then*

$$\dim_{\mathbb{C}} \mathrm{Hom}_{(\mathfrak{g},K)}(\pi, C_\eta^\infty(R\backslash G)) \leq 2.$$

We have moreover,

THEOREM 8.15. *We assume that* $\frac{d+m}{2} \in \mathbb{Z}$*,* $d \geq m > 0$ *and* $\xi = \mathrm{diag}(c_1, c_2) > 0$*. Then,*

$$\dim_{\mathbb{C}} \mathrm{Hom}_{(\mathfrak{g},K)}(\pi, C_\eta^\infty(R\backslash G)^{\mathrm{mod}}) \leq 1.$$

The generalized Whittaker functions $\Phi_{\pi,\tau}(a)$ *for the representation* $\pi = \mathrm{Ind}_{P_J}^G((\chi_m, D_p^-) \otimes e^{\nu+\rho_J} \otimes 1_{N_J})$ *with corner* K*-type* τ^**,* $\tau = [m, 0; 2p-m]$ *are described as follows.*
Let $\Phi_{\pi,\tau} = \sum_{k=0}^m c_k(a)(v_{\frac{d+m}{2}-k} \otimes f_{k0}) \in C_{M_0}^\infty(A, V_\chi \otimes W_\tau)$*, if* $\Phi_{\pi,\tau}$ *is slowly increasing when each* a_1, a_2 *tends to infinity, then* $\Phi_{\pi,\tau}$ *is up to constant uniquely determined and the coefficient functions* $c_k(a)$ *are given by*

$$\begin{aligned} c_k(a) =& (-1)^k \binom{m}{k} \binom{d}{\frac{d-m}{2}+k} (\sqrt{c_1}a_1)^{m-k+p+2} (\sqrt{c_2}a_2)^{k+p+2} \\ &\times (c_1 a_1^2 - c_2 a_2^2)^{\frac{d-m}{2}} e^{-2\pi(c_1 a_1^2 + c_2 a_2^2)} \\ &\times \int_0^1 F(2\pi c_1 a_1^2 t + 2\pi c_2 a_2^2 (1-t)) t^{\frac{d+m}{2}-k} (1-t)^{\frac{d-m}{2}+k} \, dt, \end{aligned}$$

where

$$F(x) = e^x x^{-\frac{d+p+2}{2}} W_{\frac{p-d-2}{2}, \frac{\nu}{2}}(2x),$$

with $W_{\kappa,\mu}(z)$*, classical Whittaker's confluent hypergeometric function.*

Remark. The next section tells us that this formula is also true when $m = 0$.

8.6. The case for $(D_p^-, m = 0)$. We calculate the differential equations for $\Phi_{\pi,\tau}$, for $\pi = \pi_J = \mathrm{Ind}_{P_J}^G((\chi_0, D_p^-) \otimes e^{\nu+\rho_J} \otimes 1_{N_J})$ with corner K-type τ^*, $\tau = [0,0;2p]$.

PROPOSITION 8.16. *Let* $\Phi_{\pi,\tau}(a) = c(a)(v_{\frac{d}{2}} \otimes f_{00}) \in C_M^\infty(A, V_\chi \otimes W_\tau)$*. Here,* $\eta = \chi \otimes \xi$*,* χ *is an irreducible finite-dimensional representation of* $SU(\xi)$ *and* ξ *is a nondegenerate unitary character of* N_S *determined by the positive definite matrix* $\mathrm{diag}(c_1, c_2)$*.* $\dim \chi = d+1$ *and* d *is even. Then the function* $c(a_1, a_2)$ *satisfy the following differential equations,*

$$\Big[\{\partial_1 + 4\pi c_1 a_1^2 + \frac{4c_1a_1^2}{\triangle} - (p+4)\}\{\partial_2 + 4\pi c_2 a_2^2 - p\}$$
$$+\{\partial_2 + 4\pi c_2 a_2^2 - \frac{4c_2a_2^2}{\triangle} - (p+4)\}\{\partial_1 + 4\pi c_1 a_1^2 - p\}$$
$$+4c_1c_2 d(\frac{d}{2}+1)(\frac{a_1a_2}{\triangle})^2\Big] c(a) = 0, \tag{72}$$

and

$$\Big[\partial_1^2 + \partial_2^2 - 2(\partial_1+\partial_2) + 4\frac{c_2a_2^2}{\triangle}\partial_1 - 4\frac{c_1a_1^2}{\triangle}\partial_2 - 16\pi^2(c_1^2a_1^4 + c_2^2a_2^4)$$
$$+8\pi(c_1a_1^2 + c_2a_2^2)p - 4c_1c_2d(\frac{d}{2}+1)(\frac{a_1a_2}{\triangle})^2\Big] c(a)$$
$$= \{\nu^2 + (p-1)^2 - 10\}\, c(a). \tag{73}$$

PROOF. From Proposition 8.3, we have

$$\bar{P}^{(-,-)} \circ \mathrm{Rad}(\nabla^-) \circ \bar{P}^{(+,+)} \circ \mathrm{Rad}(\nabla^-)\, \Phi_{\pi,\tau} = 0.$$

We note that

$$\tau \otimes \mathrm{Ad}^- = [0,0:2p] \otimes [1,1:-2] \overset{\bar{P}^{(+,+)}}{\longrightarrow} [1,1;2p-2],$$

and set $\tau' = [1,1;2p-2]$,

$$\tau' \otimes \mathrm{Ad}^- = [1,1:2p-2] \otimes [1,1:-2] \overset{\bar{P}^{(-,-)}}{\longrightarrow} \tau'' = [0,0;2p-4].$$

Firstly we calculate $\bar{P}^{(+,+)} \circ \mathrm{Rad}(\nabla^-)\Phi_{\pi,\tau}(a)$.

$$\bar{P}^{(+,+)} \circ \mathrm{Rad}(\nabla^-) c(a)(v_{\frac{d}{2}} \otimes f_{00})$$
$$= \Big[\frac{1}{2}\partial_1 + 2\pi c_1 a_1^2 - \frac{1}{2}\tau'(-\sqrt{-1}H_{13}) - 3 + \frac{2c_1a_1^2}{\triangle}\Big](c(a)v_{\frac{d}{2}} \otimes f_{01})$$
$$+\Big[\frac{1}{2}\frac{a_1a_2}{\triangle}\chi(X_5 + \sqrt{-1}X_6) + \tau'(e_+^2)$$

$$-\frac{c_1a_1^2}{\triangle}\tau'(e_+^1+e_+^2)\Bigg]\,(c(a)v_{\frac{d}{2}}\otimes f_{00})$$

$$+\left[\frac{1}{2}\frac{a_1a_2}{\triangle}\chi(X_5-\sqrt{-1}X_6)-\tau'(e_-^1)\right.$$

$$\left.+\frac{c_1a_1^2}{\triangle}\tau'(e_-^1+e_-^2)\right](c(a)v_{\frac{d}{2}}\otimes(-f_{11}))$$

$$+\left[\frac{1}{2}\partial_2+2\pi c_2a_2^2-\frac{1}{2}\tau'(-\sqrt{-1}H_{24})-1-\frac{2c_1a_1^2}{\triangle}\right](c(a)v_{\frac{d}{2}}\otimes(-f_{10}))$$

Here, we use the fact (from Lemma 4.4):

$$\bar{P}^{(+,+)}(f_{00}\otimes X_{31})=f_{01},\qquad \bar{P}^{(+,+)}(f_{00}\otimes X_{41})=f_{00},$$
$$\bar{P}^{(+,+)}(f_{00}\otimes X_{32})=-f_{11},\qquad \bar{P}^{(+,+)}(f_{00}\otimes X_{42})=-f_{10}.$$

Next we use the formulas from Lemmas 4.1 and 5.1:

$$\tau'(-\sqrt{-1}H_{13})(f_{01})=(p-2)f_{01},$$
$$\tau'(-\sqrt{-1}H_{24})(f_{10})=(p-2)f_{10}.$$

and

$$\frac{1}{2}\frac{a_1a_2}{\triangle}\chi(X_5+\sqrt{-1}X_6)v_{\frac{d}{2}}=-\sqrt{c_1c_2}\frac{a_1a_2}{\triangle}\frac{d}{2}v_{\frac{d}{2}+1},$$
$$\frac{1}{2}\frac{a_1a_2}{\triangle}\chi(X_5-\sqrt{-1}X_6)v_{\frac{d}{2}}=\sqrt{c_1c_2}\frac{a_1a_2}{\triangle}\frac{d}{2}v_{\frac{d}{2}-1}.$$

Then we have:

$$\bar{P}^{(+,+)}\circ\mathrm{Rad}(\nabla^-)c(a)(v_{\frac{d}{2}}\otimes f_{00})$$

$$=\left[\frac{1}{2}\partial_1+2\pi c_1a_1^2-\frac{1}{2}(p-2)-3+\frac{2c_1a_1^2}{\triangle}\right](c(a)v_{\frac{d}{2}}\otimes f_{01})$$

$$+\left[-\sqrt{c_1c_2}\frac{a_1a_2}{\triangle}\chi(X_+)+\tau'(e_+^2)-\frac{c_1a_1^2}{\triangle}\tau'(e_+^1+e_+^2)\right](c(a)v_{\frac{d}{2}}\otimes f_{00})$$

$$-\left[\sqrt{c_1c_2}\frac{a_1a_2}{\triangle}\chi(X_-)-\tau'(e_-^1)+\frac{c_1a_1^2}{\triangle}\tau'(e_-^1+e_-^2)\right](c(a)v_{\frac{d}{2}}\otimes f_{11})$$

$$-\left[\frac{1}{2}\partial_2+2\pi c_2a_2^2-\frac{1}{2}(p-2)-1-\frac{2c_1a_1^2}{\triangle}\right](c(a)v_{\frac{d}{2}}\otimes f_{10})$$

$$=\left[\frac{1}{2}\partial_1+2\pi c_1a_1^2-\frac{1}{2}(p-2)-3+\frac{2c_1a_1^2}{\triangle}\right]c(a)(v_{\frac{d}{2}}\otimes f_{01})$$

$$-\sqrt{c_1c_2}\frac{a_1a_2}{\triangle}\frac{d}{2}\,c(a)(v_{\frac{d}{2}+1}\otimes f_{00})+\left(1-\frac{c_1a_1^2}{\triangle}\right)c(a)(v_{\frac{d}{2}}\otimes f_{01})$$

$$
\begin{aligned}
&-\frac{c_1a_1^2}{\triangle}\,c(a)(v_{\frac{d}{2}}\otimes f_{10})\\
&-\sqrt{c_1c_2}\frac{a_1a_2}{\triangle}\frac{d}{2}\,c(a)(v_{\frac{d}{2}-1}\otimes f_{11})-\left(-1+\frac{c_1a_1^2}{\triangle}\right)\,c(a)(v_{\frac{d}{2}}\otimes f_{01})\\
&-\frac{c_1a_1^2}{\triangle}\,c(a)(v_{\frac{d}{2}}\otimes f_{10})\\
&-\left[\frac{1}{2}\partial_2+2\pi c_2a_2^2-\frac{1}{2}(p-2)-1-\frac{2c_1a_1^2}{\triangle}\right]\,c(a)(v_{\frac{d}{2}}\otimes f_{10})
\end{aligned}
$$

Therefore if we write the last formula as

$$
\begin{aligned}
\psi(a)=&\,b_{\frac{d}{2},10}(a)(v_{\frac{d}{2}}\otimes f_{10})+b_{\frac{d}{2}-1,11}(a)(v_{\frac{d}{2}-1}\otimes f_{11})\\
&+b_{\frac{d}{2}+1,00}(a)(v_{\frac{d}{2}+1}\otimes f_{00})+b_{\frac{d}{2},01}(a)(v_{\frac{d}{2}+1}\otimes f_{01}).
\end{aligned}
$$

Then these functions are given by

$$
b_{\frac{d}{2},10}(a)=-\left[\frac{1}{2}\partial_2+2\pi c_2a_2^2-\frac{1}{2}p\right]\,c(a), \tag{74}
$$

$$
b_{\frac{d}{2}-1,11}(a)=b_{\frac{d}{2}+1,00}(a)=-\frac{d}{2}\sqrt{c_1c_2}\frac{a_1a_2}{\triangle}\,c(a), \tag{75}
$$

$$
b_{\frac{d}{2},01}(a)=\left[\frac{1}{2}\partial_1+2\pi c_1a_1^2-\frac{1}{2}p\right]\,c(a). \tag{76}
$$

Next we calculate $\bar{P}^{(-,-)}\circ \mathrm{Rad}(\nabla^-)\psi(a)$.

$$
\begin{aligned}
&\bar{P}^{(-,-)}\circ \mathrm{Rad}(\nabla^-)\psi(a)\\
&=\left[\frac{1}{2}\partial_1+2\pi c_1a_1^2-\frac{1}{2}\tau''(-\sqrt{-1}H_{13})-3\right.\\
&\qquad\left.+\frac{2c_1a_1^2}{\triangle}\right]\,(b_{\frac{d}{2},10}(a)v_{\frac{d}{2}}\otimes(-f_{00}))\\
&\quad+\left[\frac{1}{2}\frac{a_1a_2}{\triangle}\chi(X_5+\sqrt{-1}X_6)+\tau''(e_+^2)\right.\\
&\qquad\left.-\frac{c_1a_1^2}{\triangle}\tau''(e_+^1+e_+^2)\right]\,(b_{\frac{d}{2}-1,11}(a)v_{\frac{d}{2}}\otimes f_{00})\\
&\quad+\left[\frac{1}{2}\frac{a_1a_2}{\triangle}\chi(X_5-\sqrt{-1}X_6)-\tau''(e_-^1)\right.\\
&\qquad\left.+\frac{c_1a_1^2}{\triangle}\tau''(e_-^1+e_-^2)\right]\,(b_{\frac{d}{2}+1,00}(a)v_{\frac{d}{2}+1}\otimes(-f_{00}))\\
&\quad+\left[\frac{1}{2}\partial_2+2\pi c_2a_2^2-\frac{1}{2}\tau'(-\sqrt{-1}H_{24})-1\right.
\end{aligned}
$$

$$-\frac{2c_1a_1^2}{\triangle}\Big] (b_{\frac{d}{2},01}(a)v_{\frac{d}{2}} \otimes (-f_{10}))$$

Here, we use the fact (from Lemma 4.4):

$$\bar{P}^{(-,-)}(f_{10} \otimes X_{31}) = -f_{00}, \qquad \bar{P}^{(-,-)}(f_{11} \otimes X_{41}) = f_{00},$$
$$\bar{P}^{(-,-)}(f_{00} \otimes X_{32}) = -f_{00}, \qquad \bar{P}^{(-,-)}(f_{01} \otimes X_{42}) = f_{00}.$$

Next we use the formula from Lemmas 4.1 and 5.1:

$$\tau''(-\sqrt{-1}H_{13})(f_{00}) = (p-2)f_{00},$$
$$\tau''(-\sqrt{-1}H_{24})(f_{00}) = (p-2)f_{00}.$$

and

$$\frac{1}{2}\frac{a_1a_2}{\triangle}\chi(X_5 + \sqrt{-1}X_6)v_{\frac{d}{2}-1} = -\sqrt{c_1c_2}\frac{a_1a_2}{\triangle}(\frac{d}{2}+1)v_{\frac{d}{2}},$$
$$\frac{1}{2}\frac{a_1a_2}{\triangle}\chi(X_5 - \sqrt{-1}X_6)v_{\frac{d}{2}+1} = \sqrt{c_1c_2}\frac{a_1a_2}{\triangle}(\frac{d}{2}+1)v_{\frac{d}{2}}.$$

Then we have:

$$\begin{aligned}
&\bar{P}^{(-,-)} \circ \mathrm{Rad}(\nabla^-)\psi(a) \\
&= -\left[\frac{1}{2}\partial_1 + 2\pi c_1 a_1^2 - \frac{1}{2}(p-2) - 3 + \frac{2c_1a_1^2}{\triangle}\right] b_{\frac{d}{2},10}(a)(v_{\frac{d}{2}} \otimes f_{00}) \\
&\quad -(\frac{d}{2}+1)\sqrt{c_1c_2}\frac{a_1a_2}{\triangle} b_{\frac{d}{2}-1,11}(a)(v_{\frac{d}{2}} \otimes f_{00}) \\
&\quad -(\frac{d}{2}+1)\sqrt{c_1c_2}\frac{a_1a_2}{\triangle} b_{\frac{d}{2}+1,00}(a)(v_{\frac{d}{2}} \otimes f_{00}) \\
&\quad +\left[\frac{1}{2}\partial_2 + 2\pi c_2 a_2^2 - \frac{1}{2}(p-2) - 1 - \frac{2c_1a_1^2}{\triangle}\right] b_{\frac{d}{2},01}(a)(v_{\frac{d}{2}} \otimes f_{00}) = 0.
\end{aligned}$$

Thus we have

$$\begin{aligned}
&-\left[\frac{1}{2}\partial_1 + 2\pi c_1 a_1^2 - \frac{1}{2}(p-2) - 3 + \frac{2c_1a_1^2}{\triangle}\right] b_{\frac{d}{2},10}(a) \\
&+\left[\frac{1}{2}\partial_2 + 2\pi c_2 a_2^2 - \frac{1}{2}(p-2) - 1 - \frac{2c_1a_1^2}{\triangle}\right] b_{\frac{d}{2},01}(a) \\
&-(\frac{d}{2}+1)\sqrt{c_1c_2}\frac{a_1a_2}{\triangle}\left\{b_{\frac{d}{2}-1,11}(a) + b_{\frac{d}{2}+1,00}(a)\right\} = 0. \qquad (77)
\end{aligned}$$

This is the desired (72). The equation (73) is obtained if we put $m = k = 0$ and $j = d/2$ in (51). $\square$

PROPOSITION 8.17. *Under the same assumptions and notations in the above Proposition, if we write* $c(a) = (a_1a_2)^{p+2}e^{-2\pi(c_1a_1^2+c_2a_2^2)}h(a)$, *then the function* $h(a)$ *satisfy the following differential equations,*

$$\left[\partial_1\partial_2 - \frac{2c_2a_2^2}{\triangle}\partial_1 + \frac{2c_1a_1^2}{\triangle}\partial_2 + 4c_1c_2\left(\frac{a_1a_2}{\triangle}\right)^2\frac{d}{2}\left(\frac{d}{2}+1\right)\right] h(a) = 0. \quad (78)$$

and

$$\Big[(\partial_1+\partial_2)^2 + 2(p+1)(\partial_1+\partial_2) - 8\pi(c_1a_1^2\partial_1 + c_2a_2^2\partial_2)$$
$$-16\pi(c_1a_1^2 + c_2a_2^2) + (p+1)^2 - \nu^2\Big]h(a) = 0. \quad (79)$$

PROOF. We rewrite (72) by $h(a)$, then

$$\Big[\{\partial_1 + \frac{4c_1a_1^2}{\triangle} - 2\}\{\partial_2 + 2\} + \{\partial_2 - \frac{4c_2a_2^2}{\triangle} - 2\}\{\partial_1 + 2\}$$
$$+4c_1c_2d\left(\frac{d}{2}+1\right)\left(\frac{a_1a_2}{\triangle}\right)^2\Big] h(a) = 0.$$

Thus we have

$$\left[\partial_1\partial_2 - \frac{2c_2a_2^2}{\triangle}\partial_1 + \frac{2c_1a_1^2}{\triangle}\partial_2 + 4c_1c_2\left(\frac{a_1a_2}{\triangle}\right)^2\frac{d}{2}\left(\frac{d}{2}+1\right)\right] h(a) = 0. \quad (80)$$

□

Two partial differential equations obtained above are same equations which is obtained in the case $(D_p^-, m > 0)$ and put $m = 0$ formally. Therefore analytic solutions are already studied.

CHAPTER 4

Generalized Whittaker functions for the discrete series representations

9. Generalized Whittaker functions for the discrete series representations

9.1. Parametrization of the discrete series representations. Let us parametrize the discrete series representations of $SU(2,2)$. There are exactly six positive systems of the absolute roots, $\Delta_{\rm I}^+$, $\Delta_{\rm II}^+$, ..., $\Delta_{\rm VI}^+$ containing compact positive roots Δ_c^+.

$$\begin{aligned}
\Delta_{\rm I}^+ &= \{[2,0;0],[1,-1;2],[1,1;2],[-1,-1;2],[-1,1;2],[0,2;0]\},\\
\Delta_{\rm II}^+ &= \{[1,-1;2],[2,0;0],[1,1;2],[1,1;-2],[0,2;0],[-1,1;2]\},\\
\Delta_{\rm III}^+ &= \{[1,-1;2],[1,1;2],[2,0;0],[0,2;0],[1,1;-2],[1,-1;-2]\},\\
\Delta_{\rm IV}^+ &= \{[-1,1;-2],[1,1;-2],[0,2;0],[2,0;0],[1,1;2],[-1,1;2]\},\\
\Delta_{\rm V}^+ &= \{[-1,1;-2],[0,2;0],[1,1;-2],[1,1;2],[2,0;0],[1,-1;-2]\},\\
\Delta_{\rm VI}^+ &= \{[0,2;0],[-1,1;-2],[1,1;-2],[-1,-1;-2],[1,-1;-2],[2,0;0]\}.
\end{aligned}$$

For each $J \in \{{\rm I},{\rm II},{\rm III},{\rm IV},{\rm V},{\rm VI}\}$, we write $\Delta_{J,n}^+ = \Delta_J^+ \setminus \Delta_c^+$ the set of non-compact positive roots. Then we define the subset of dominant weights by

$$\Xi_J = \Big\{\Lambda = [\Lambda_1,\Lambda_2;\Lambda_3] | \Lambda \text{ is } \Delta_{J,n}^+\text{-dominant}\Big\}.$$

The space $\coprod_{J={\rm I}}^{\rm VI} \Xi_J$ gives the Harish-Chandra parametrization of the discrete series representations for $SU(2,2)$.

Let π_Λ be the discrete series representation of G with the Harish-Chandra parameter $\Lambda \in \Xi_J$. The Blattner parameter of π_Λ is given by $\lambda = \Lambda - \rho_c + \rho_{J,n}$, where ρ_c is the half sum of compact positive roots and $\rho_{J,n}$ is the half sum of non-compact positive roots in $\Delta_{J,n}^+$ respectively. In particular the Blattner formula says that π_Λ has the minimal K-type whose highest weight is λ. For each J, λ is given as,

$$\lambda = \begin{cases} \Lambda + [-1,-1;4] & \text{for} \quad \Lambda \in \Xi_{\mathrm{I}}, \\ \Lambda + [0,0;2] & \text{for} \quad \Lambda \in \Xi_{\mathrm{II}}, \\ \Lambda + [1,-1;0] & \text{for} \quad \Lambda \in \Xi_{\mathrm{III}}, \\ \Lambda + [-1,1;0] & \text{for} \quad \Lambda \in \Xi_{\mathrm{IV}}, \\ \Lambda + [0,0;-2] & \text{for} \quad \Lambda \in \Xi_{\mathrm{V}}, \\ \Lambda + [-1,-1;-4] & \text{for} \quad \Lambda \in \Xi_{\mathrm{VI}}. \end{cases} \tag{81}$$

As is determined in [**Ya2**], the Gelfand-Kirillov dimension of the π_Λ are given as follows:

$$\text{GK-}\dim(\pi_\Lambda) = \begin{cases} 4 & \text{for} \quad \Lambda \in \Xi_{\mathrm{I}} \cup \Xi_{\mathrm{VI}}, \\ 6 & \text{for} \quad \Lambda \in \Xi_{\mathrm{II}} \cup \Xi_{\mathrm{V}}, \\ 5 & \text{for} \quad \Lambda \in \Xi_{\mathrm{III}} \cup \Xi_{\mathrm{IV}}. \end{cases} \tag{82}$$

Therefore the representations with the Harish-Chandra parameter $\Lambda \in \Xi_{\mathrm{II}} \cup \Xi_{\mathrm{V}}$ are called the large discrete series in the sense of Vogan [**Vog**], and has an ordinary Whittaker model. The representations with the Harish-Chandra parameter $\Lambda \in \Xi_{\mathrm{I}}$ (resp. Ξ_{VI}) is called the holomorphic (resp. anti-holomorphic) discrete series. We call the discrete series representations whose Gelfand-Kirillov dimension is 5, the middle discrete series.

9.2. Characterization of generalized Whittaker functions. Using the Schmid operator ∇_{η,τ_μ}, we define the following projectors

$$P_{\tau_\lambda}^{(J)} : W_\lambda \otimes \mathfrak{p}_{\mathbb{C}} \to W_\lambda^- = \bigoplus_{\beta \in \Delta_{J,n}^+} W_{\lambda-\beta}. \tag{83}$$

THEOREM 9.1 (Yamashita [**Ya2**]). *Let π_Λ be the discrete series representation of G with a Harish-Chandra parameter $\Lambda \in \Xi_J$. Assume that the Blattner parameter λ of π_Λ satisfies the condition (FFW) which stated below. Then,*

$$\mathrm{Hom}_{(\mathfrak{g},K)}(\pi_\Lambda^*, C_\eta^\infty(R\backslash G)) \simeq \ker(P_{\tau_\lambda}^{(J)} \circ \nabla_{\eta,\tau_\lambda}).$$

(FFW): $\lambda - \sum_{\beta \in Q}$ is Δ_c^+-dominant for any subset Q of $\Delta_{J,n}^+$, i.e., λ is far from the walls.

In the succeeding sections, by using the above theorem we write down the partial differential equations which characterize the generalized Whittaker functions more concretely for π_Λ, $\Lambda \in \Xi_J$ with $J = \mathrm{I}, \mathrm{II}, \mathrm{III}$. Next we studies analytic solutions for the obtained system of partial differential equations for each cases respectively.

Remark. The original theorem required the condition: (FFW). But it is known that this condition should be deleted by an argument using "translation principle".

10. Generalized Whittaker functions for the holomorphic discrete series representations

Generalized Whittaker functions for the holomorphic discrete series representations appear in the Fourier expansion of holomorphic Hermitian cusp forms. They are elementary functions and well known. So we state the results without proof.

PROPOSITION 10.1. *Generalized Whittaker functions $\Phi_{\pi,\tau}$ for $\pi = \pi^*_\Lambda$, $\Lambda \in \Xi_{\mathrm{I}}$, with their minimal K-type τ^*, $\tau = [r,s;u]$ satisfy the following equations.*

$$P^{(\mathrm{I})}\nabla\Phi_{\pi,\tau}(a) = 0. \tag{84}$$

Here, $P^{(\mathrm{I})} = \bar{P}^{(-,-)} \oplus \bar{P}^{(-,+)} \oplus \bar{P}^{(+,-)} \oplus \bar{P}^{(+,+)}$, equivalent to

$$\mathrm{Rad}(\nabla^-)\Phi_{\pi,\tau}(a) = 0. \tag{85}$$

PROOF. This is a direct consequence of Theorem 9.1.

$$\Delta^+_{\mathrm{I}} = \{[2,0;0],[1,-1;2],[1,1:2],[-1,-1;2],[-1,1;2],[0,2;0]\}$$

and that

$$\Delta^+_{\mathrm{I},n} = \{[1,-1;2],[1,1:2],[-1,-1;2],[-1,1;2]\}.$$

or by Cauchy-Riemann condition. □

We write down the system of differential equations of generalized Whittaker functions $\Phi_{\pi,\tau}(a)$ for the case $\pi = \pi^*_\Lambda$, $\Lambda \in \Xi_{\mathrm{I}}$ with their minimal K-type τ^*, $\tau = [r,s;u]$ and solve it.

Let $\Phi_{\pi,\tau}(a) = \sum_{j=0}^{d} \sum_{\substack{0\le k\le r \\ 0\le l\le s}} c_{jkl}(a)(v_j \otimes f_{kl}) \in C^\infty_{M_0}(A; V_\chi \otimes W_\tau)$. Here, $\eta = \chi \otimes \xi$, χ is an irreducible finite dimensional representation of $SU(\xi) \simeq SU(2)$ and ξ is a non-trivial unitary character of N_S determined by a positive definite matrix $H_\xi = \begin{pmatrix} c_1 & \\ & c_2 \end{pmatrix}$. We assume that $\dim\chi = d+1$, $\frac{d+r+s}{2} \in \mathbb{Z}$.

Then we have,

THEOREM 10.2. *If $d \neq r+s$ then*

$$\dim_{\mathbb{C}} \mathrm{Hom}_{(\mathfrak{g},K)}(\pi^*_\Lambda, C^\infty_\eta(R\backslash G)) = 0,$$

and if $d = r+s$ then

$$\dim_{\mathbb{C}} \mathrm{Hom}_{(\mathfrak{g},K)}(\pi^*_\Lambda, C^\infty_\eta(R\backslash G)) = 1.$$

We assume that $d = r+s$ and $\xi = \mathrm{diag}(c_1,c_2) > 0$. Then,

$$\dim_{\mathbb{C}} \mathrm{Hom}_{(\mathfrak{g},K)}(\pi^*_\Lambda, C^\infty_\eta(R\backslash G)^{\mathrm{mod}}) = 1.$$

The generalized Whittaker functions for the holomorphic discrete series representations for $\pi = \pi^*_\Lambda$, $\Lambda \in \Xi_{\mathrm{I}}$ with minimal K-type are described

as follows.
Let $\Phi_{\pi,\tau}(a) = \sum_{\substack{0\le k\le r \\ 0\le l\le s}} c_{kl}(a)(v_{r+s-k-l}\otimes f_{kl}) \in C^\infty_{M_0}(A; V_\chi \otimes W_\tau)$, *then* $\Phi_{\pi,\tau}(a)$ *are up to constant uniquely determined and the coefficient functions are given as*

$$c_{kl}(a) = (-1)^{k+l}\binom{r}{k}\binom{s}{l}(\sqrt{c_1}a_1)^{b_1+k-l}(\sqrt{c_2}a_2)^{b_2-k+l}e^{-2\pi(c_1a_1^2+c_2a_2^2)}. \tag{86}$$

Here, $b_1 = (-r+s+u)/2$ *and* $b_2 = (r-s+u)/2$.

11. Generalized Whittaker functions for the large discrete series representations

11.1. Large discrete series representations.

PROPOSITION 11.1. *Generalized Whittaker functions $\Phi_{\pi,\tau}$ for $\pi = \pi^*_\Lambda$, $\Lambda \in \Xi_{\rm II}$, with their minimal K-type τ^*, $\tau = [r,s;u]$ satisfy the following equations.*

$$P^{(\rm II)}\nabla\Phi_{\pi,\tau}(a) = 0. \tag{87}$$

Here, $P^{(\rm II)} = P^{(-,-)} \oplus \bar{P}^{(-,-)} \oplus \bar{P}^{(-,+)} \oplus \bar{P}^{(+,-)}$.

PROOF. This is a direct consequence of Theorem 9.1.

$$\Delta^+_{\rm II} = \{[1,-1;2],[2,0;0],[1,1:2],[1,1;-2],[0,2;0],[-1,1;2]\}$$

and that

$$\Delta^+_{{\rm II},n} = \{[1,-1;2],[1,1:2],[1,1;-2],[-1,1;2]\}.$$

□

11.2. Explicit differential equations. We write down explicit differential equations of generalized Whittaker functions $\Phi_{\pi,\tau}(a)$ for the case:
$\lceil\ \pi = \pi^*_\Lambda$, $\Lambda \in \Xi_{\rm II}$ with their minimal K-type τ^*, $\tau = [r,s;u]\ \rfloor$.

Let $\Phi_{\pi,\tau}(a) = \sum_{j=0}^{d}\sum_{\substack{0\le k\le r\\ 0\le l\le s}} c_{jkl}(a)(v_j\otimes f_{kl}) \in C^\infty_{M_0}(A;V_\chi\otimes W_\tau)$. Here, $\eta = \chi\otimes\xi$, χ is an irreducible finite-dimensional representation of $SU(\xi)\simeq SU(2)$ and ξ is a non-trivial unitary character of N_S determined by a positive definite matrix $H_\xi = \begin{pmatrix} c_1 & \\ & c_2\end{pmatrix}$. We assume that $\dim\chi = d+1$, $\frac{d+r+s}{2}\in\mathbb{Z}$.

Then the unknown functions $c_{jkl}(a)$ satisfy the following difference-differential equations.

PROPOSITION 11.2. *Under the above assumptions the coefficient functions $c_{jkl}(a)$ in $\Phi_{\pi,\tau}(a)$ satisfy the following system of partial differential equations. Here, we put $\partial_i = a_i\frac{\partial}{\partial a_i}$: the Euler operator $(i=1,2)$ and $\triangle = (c_1a_1^2 - c_2a_2^2)$.*

$$\begin{aligned}&(P^{(-,-)})_{jkl}\\ &\quad -\sqrt{c_1c_2}\frac{a_1a_2}{\triangle}(d-j+1)(k+1)(l+1)\,c_{j-1,k+1,l+1}(a)\\ &\quad +\sqrt{c_1c_2}\frac{a_1a_2}{\triangle}(j+1)(k-r)(s-l)\,c_{j+1,k,l}(a)\end{aligned}$$

$$
\begin{aligned}
&+\frac{1}{2}(k-r)(l+1)\Big[\partial_1 - 4\pi c_1 a_1^2 + (b_1 - k + l - 2s - 3) \\
&\qquad\qquad +(s-l+k+1)\frac{2c_1a_1^2}{\triangle}\Big]\; c_{j,k,l+1}(a) \\
&+\frac{1}{2}(k+1)(s-l)\Big[\partial_2 - 4\pi c_2 a_2^2 + (b_2 + k - 2r - l - 3) \\
&\qquad\qquad -(r-k+l+1)\frac{2c_2a_2^2}{\triangle}\Big]\; c_{j,k+1,l}(a) \\
&= 0 \qquad (0 \le j \le d,\ 0 \le k \le r-1,\ 0 \le l \le s-1),
\end{aligned}
$$

$$
\begin{aligned}
&(\bar{P}^{(-,+)})_{jkl} \\
&\quad \sqrt{c_1c_2}\frac{a_1a_2}{\triangle}(d-j+1)(k+1)\; c_{j-1,k+1,l}(a) \\
&\quad +\sqrt{c_1c_2}\frac{a_1a_2}{\triangle}(j+1)(k-r)\; c_{j+1,k,l-1}(a) \\
&\quad -\frac{1}{2}(k+1)\Big[\partial_1 + 4\pi c_1 a_1^2 - (b_1 - k + l + 2r - 2s) \\
&\qquad\qquad +(r-k-s+l-1)\frac{2c_1a_1^2}{\triangle}\Big]\; c_{j,k+1,l-1}(a) \\
&\quad +\frac{1}{2}(k-r)\Big[\partial_2 + 4\pi c_2 a_2^2 - (b_2 + k - l + 2) \\
&\qquad\qquad -(k-l+1)\frac{2c_2a_2^2}{\triangle}\Big]\; c_{j,k,l}(a) \\
&\quad = 0 \qquad (0 \le j \le d,\ 0 \le k \le r-1,\ 0 \le l \le s+1),
\end{aligned}
$$

$$
\begin{aligned}
&(\bar{P}^{(-,-)})_{jkl} \\
&\quad -\sqrt{c_1c_2}\frac{a_1a_2}{\triangle}(d-j+1)(k+1)(l+1)\; c_{j-1,k+1,l+1}(a) \\
&\quad +\sqrt{c_1c_2}\frac{a_1a_2}{\triangle}(j+1)(k-r)(s-l)\; c_{j+1,k,l}(a) \\
&\quad +\frac{1}{2}(k+1)(l-s)\Big[\partial_1 + 4\pi c_1 a_1^2 - (b_1 - k + l + 2r + 3) \\
&\qquad\qquad +(r-k+l+1)\frac{2c_1a_1^2}{\triangle}\Big]\; c_{j,k+1,l}(a) \\
&\quad +\frac{1}{2}(r-k)(l+1)\Big[\partial_2 + 4\pi c_2 a_2^2 - (b_2 + k - l + 2s + 3) \\
&\qquad\qquad -(k+s-l+1)\frac{2c_2a_2^2}{\triangle}\Big]\; c_{j,k,l+1}(a)
\end{aligned}
$$

$$= 0 \qquad (0 \le j \le d,\ 0 \le k \le r-1,\ 0 \le l \le s-1),$$

$$\begin{aligned}
&(\bar{P}^{(+,-)})_{jkl} \\
&\quad \sqrt{c_1c_2}\frac{a_1a_2}{\triangle}(d-j+1)(l+1)\, c_{j-1,k,l+1}(a) \\
&\quad +\sqrt{c_1c_2}\frac{a_1a_2}{\triangle}(j+1)(l-s)\, c_{j+1,k-1,l}(a) \\
&\quad +\frac{1}{2}(s-l)\Big[\partial_1 + 4\pi c_1 a_1^2 - (b_1 - k + l + 2) \\
&\qquad\qquad +(1+l-k)\frac{2c_1a_1^2}{\triangle}\Big]\, c_{j,k,l}(a) \\
&\quad +\frac{1}{2}(l+1)\Big[\partial_2 + 4\pi c_2 a_2^2 - (b_2 + k - l - 2r + 2s) \\
&\qquad\qquad -(k-l-r+s-1)\frac{2c_2a_2^2}{\triangle}\Big]\, c_{j,k-1,l+1}(a) \\
&\quad = 0 \qquad (0 \le j \le d,\ 0 \le k \le r+1,\ 0 \le l \le s-1).
\end{aligned}$$

PROOF. Proposition 11.1 says $P \circ \mathrm{Rad}(\nabla)\Phi_{\pi,\tau}(a) = 0$, for $P = P^{(-,-)}$, $\bar{P}^{(-,+)}$, $\bar{P}^{(-,-)}$ and $\bar{P}^{(+,-)}$. For example, we write down concretely for $P = P^{(-,-)}$,

$$\begin{aligned}
&P^{(-,-)} \circ \mathrm{Rad}(\nabla^+)\Phi(a) \\
&\quad = \sum_{j,k,l} P^{(-,-)} \circ \mathrm{Rad}(\nabla^+) c_{jkl}(a)(v_j \otimes f_{kl}) \\
&\quad = \sum_{j,k,l} [A^{(-,-)}_{jkl}(a)(v_{j+1} \otimes f_{k-1,l-1}) + B^{(-,-)}_{jkl}(a)(v_{j-1} \otimes f_{k,l}) \\
&\qquad\qquad + C^{(-,-)}_{jkl}(a)(v_j \otimes f_{k,l-1}) + D^{(-,-)}_{jkl}(a)(v_j \otimes f_{k-1,l})],
\end{aligned}$$

where these 4 functions $A^{(-,-)}_{jkl}(a), \dots, D^{(-,-)}_{jkl}(a)$ are defined in Lemma 7.1 in Section 7 for the radial parts of the shift operators. So $P^{(-,-)} \circ \mathrm{Rad}(\nabla^+)\Phi(a) = 0$ is equivalent to

$$A^{(-,-)}_{j-1,k+1,l+1}(a) + B^{(-,-)}_{j+1,k,l}(a) + C^{(-,-)}_{j,k,l+1}(a) + D^{(-,-)}_{j,k+1,l}(a) = 0. \tag{88}$$

This is precise $(P^{(-,-)})_{jkl}$. The other three difference-differential equations of $c_{jkl}(a)$ are obtained in the same manner. □

11.3. Reduction of the system of partial differential equations. We reduce the system of partial differential equations of the unknown functions $c_{jkl}(a)$ in Proposition 11.2 to a simpler holonomic system step by step.

We firstly reduce the system to the following.

PROPOSITION 11.3. *Under the same assumptions and notations in Proposition 11.2, the functions $c_{jkl}(a)$ satisfy the following system of partial differential equations. And this system is equivalent to the system in Proposition 11.2.*

$$
\begin{aligned}
&(C_1^+)_{jkl}\\
&\quad -\sqrt{c_1c_2}\frac{a_1a_2}{\triangle}(d-j+1)(k+1)(l+1)\, c_{j-1,k+1,l+1}(a)\\
&\quad +\sqrt{c_1c_2}\frac{a_1a_2}{\triangle}(j+1)(k-r)(s-l)\, c_{j+1,k,l}(a)\\
&\quad +\frac{1}{2}(k-r)(l+1)\Big[\partial_1-4\pi c_1a_1^2+(b_1-k+l-2s-3)\\
&\qquad\qquad +(s-l+k+1)\frac{2c_1a_1^2}{\triangle}\Big]\, c_{j,k,l+1}(a)\\
&\quad +\frac{1}{2}(k+1)(s-l)\Big[\partial_2-4\pi c_2a_2^2+(b_2+k-2r-l-3)\\
&\qquad\qquad -(r-k+l+1)\frac{2c_2a_2^2}{\triangle}\Big]\, c_{j,k+1,l}(a)\\
&\quad =0 \qquad (0\le j\le d,\ 0\le k\le r-1,\ 0\le l\le s-1),
\end{aligned}
$$

$$
\begin{aligned}
&(C_2^- : 1)_{jkl}\\
&\quad -\sqrt{c_1c_2}\frac{a_1a_2}{\triangle}(d-j+1)(k+1)\, c_{j-1,k+1,l+1}(a)\\
&\quad +(k+1)(l-s)\frac{c_2a_2^2}{\triangle}\, c_{j,k+1,l}(a)\\
&\quad +\frac{1}{2}(r-k)\Big[\partial_2+4\pi c_2a_2^2-(b_2+k+l+3)\\
&\qquad\qquad -(k+1)\frac{2c_2a_2^2}{\triangle}\Big]\, c_{j,k,l+1}(a)\\
&\quad =0 \qquad (0\le j\le d,\ 0\le k\le r-1,\ -1\le l\le s-1),
\end{aligned}
$$

$$
\begin{aligned}
&(C_2^- : 2)_{jkl}\\
&\quad \sqrt{c_1c_2}\frac{a_1a_2}{\triangle}(j+1)(k-r)\, c_{j+1,k,l}(a)\\
&\quad +(k-r)(l+1)\frac{c_1a_1^2}{\triangle}\, c_{j,k,l+1}(a)\\
&\quad -\frac{1}{2}(k+1)\Big[\partial_1+4\pi c_1a_1^2-(b_1-k-l+2r+1)
\end{aligned}
$$

$$
\begin{aligned}
&\quad +(r-k)\frac{2c_1a_1^2}{\triangle}\Big] \, c_{j,k+1,l}(a) \\
&= 0 \qquad (0 \le j \le d,\ 0 \le k \le r-1,\ 0 \le l \le s),
\end{aligned}
$$

$(C_3^- : 1)_{jkl}$

$$
\begin{aligned}
&\sqrt{c_1c_2}\frac{a_1a_2}{\triangle}(j+1)(l-s)\, c_{j+1,k,l}(a) \\
&+(k+1)(l-s)\frac{c_2a_2^2}{\triangle}\, c_{j,k+1,l}(a) \\
&+\frac{1}{2}(l+1)\Big[\partial_2 + 4\pi c_2a_2^2 - (b_2 - k - l + 2s + 1) \\
&\qquad +(l-s)\frac{2c_2a_2^2}{\triangle}\Big]\, c_{j,k,l+1}(a) \\
&= 0 \qquad (0 \le j \le d,\ 0 \le k \le r,\ 0 \le l \le s-1),
\end{aligned}
$$

$(C_3^- : 2)_{jkl}$

$$
\begin{aligned}
&-\sqrt{c_1c_2}\frac{a_1a_2}{\triangle}(d-j+1)(l+1)\, c_{j-1,k+1,l+1}(a) \\
&+(k-r)(l+1)\frac{c_1a_1^2}{\triangle}\, c_{j,k,l+1}(a) \\
&+\frac{1}{2}(l-s)\Big[\partial_1 + 4\pi c_1a_1^2 - (b_1 + k + l + 3) \\
&\qquad +(l+1)\frac{2c_1a_1^2}{\triangle}\Big]\, c_{j,k+1,l}(a) \\
&= 0 \qquad (0 \le j \le d,\ -1 \le k \le r-1,\ 0 \le l \le s-1),
\end{aligned}
$$

$(CS)_{jkl}$

$$
\begin{aligned}
&\Big[(\partial_1+\partial_2)^2 - (r+s+u+6)(\partial_1+\partial_2) + 4\pi(c_1a_1^2 + c_2a_2^2)(\partial_1+\partial_2) \\
&-4\pi(c_1a_1^2 - c_2a_2^2)(\partial_1-\partial_2) - 16\pi^2(c_1a_1^2 - c_2a_2^2)^2 \\
&-4\pi(r+s+4)(c_1a_1^2 + c_2a_2^2) - 4\pi(r-2k-s+2l)(c_1a_1^2 - c_2a_2^2) \\
&+(r+s+4)(u+2)\Big]\, c_{j,k,l}(a) = 0.
\end{aligned}
$$

PROOF. The equations $(C_2^- : 1)_{jkl}$ and $(C_2^- : 2)_{jkl}$ are obtained from:

$$
(\bar{P}^{(-,-)})_{j,k,l} - (s-l)(\bar{P}^{(-,+)})_{j,k,l+1}, \quad (\bar{P}^{(-,-)})_{j,k,l} + (l+1)(\bar{P}^{(-,+)})_{j,k,l+1},
$$

respectively. The equations $(C_3^- : 1)$ and $(C_3^- : 2)$ are obtained from $(\bar{P}^{-,-})_{j,k+1,l}$ and $(\bar{P}^{+,-})_{j,k+1,l}$ similarly. $(C_1^+)_{jkl} = (P^{(-,-)})_{jkl}$.

We prove $(CS)_{jkl}$. From $(P^{(-,-)})_{jkl} - (\bar{P}^{(-,-)})_{jkl}$, then

$$\begin{aligned}(k-r)(l+1)&\Big[(\partial_1+\partial_2) - 4\pi(c_1a_1^2 - c_2a_2^2) - (r+s+4)\Big]\, c_{j,k,l+1}(a)\\
+(k+1)(s-l)&\Big[(\partial_1+\partial_2) + 4\pi(c_1a_1^2 - c_2a_2^2)\\
&-(r+s+4)\Big]\, c_{j,k+1,l}(a) = 0. \end{aligned} \tag{89}$$

Next from $(C_2^- : 2)_{jkl} \times (l-s) - (C_3^- : 1)_{jkl} \times (k-r)$,

$$\begin{aligned}&(l-s)(k-r)(l+1)\frac{c_1a_1^2}{\triangle}\, c_{j,k,l+1}(a)\\
&-\frac{1}{2}(l-s)(k+1)\Big[\partial_1 + 4\pi c_1a_1^2 - (b_1-k-l+2r+1)\\
&\qquad +(r-k)\frac{2c_1a_1^2}{\triangle}\Big]\, c_{j,k+1,l}(a)\\
&-(k-r)(k+1)(l-s)\frac{c_2a_2^2}{\triangle}\, c_{j,k+1,l}(a)\\
&-\frac{1}{2}(k-r)(l+1)\Big[\partial_2 + 4\pi c_2a_2^2 - (b_2-k-l+2s+1)\\
&\qquad +(l-s)\frac{2c_2a_2^2}{\triangle}\Big]\, c_{j,k,l+1}(a) = 0.\end{aligned}$$

So we have,

$$\begin{aligned}&(k-r)(l+1)[\partial_2 + 4\pi c_2a_2^2 - (b_2-k+l+1)]\, c_{j,k,l+1}(a)\\
&= (k+1)(s-l)[\partial_1 + 4\pi c_1a_1^2 - (b_1+k-l+1)]\, c_{j,k+1,l}(a).\end{aligned} \tag{90}$$

Since two partial differential operators :

$$\begin{aligned}P_1 &= \partial_2 + 4\pi c_2a_2^2 - (b_2-k+l+1),\\
P_2 &= (\partial_1+\partial_2) - 4\pi(c_1a_1^2 - c_2a_2^2) - (r+s+4),\end{aligned}$$

are commutative, we operate P_1 on both sides of (89) and use (90). Then we have

$$\begin{aligned}&\Big[(\partial_1+\partial_2) - 4\pi(c_1a_1^2 - c_2a_2^2) - (r+s+4)\Big]\Big[\partial_1 + 4\pi c_1a_1^2\\
&-(b_1+k-l+1)\Big]\, c_{j,k+1,l}(a) + \Big[\partial_2 + 4\pi c_2a_2^2 - (b_2-k+l+1)\Big]\\
&\times\Big[(\partial_1+\partial_2) + 4\pi(c_1a_1^2 - c_2a_2^2) - (r+s+4)\Big]\, c_{j,k+1,l}(a) = 0.\end{aligned}$$

And substituting k into $k-1$, we get $(CS)_{jkl}$. □

Hereafter, we assume that $c_{jkl}(a)$ is *analytic* on $A \simeq (\mathbb{R}_{>0})^2$, and study analytic solutions for the above system of differential equations.

Firstly we reduce the above system into the following one, before stating a lemma on analyticity of $c_{jkl}(a)$. Note that this reduction has nothing to do with analyticity.

PROPOSITION 11.4. *Under the same assumptions and notations in Prop. 11.3, if we write* $c_{jkl}(a) = (\sqrt{c_1}a_1)^{b_0+r-k+l+2}(\sqrt{c_2}a_2)^{b_0+k+s-l+2} \times e^{-2\pi(c_1a_1^2+c_2a_2^2)}h_{jkl}(a)$, *then the functions* $h_{jkl}(a)$ *satisfy the following system of partial differential equations.*

$$(H_2^- : 1)_{jkl}$$

$$\begin{aligned}
&-\frac{c_2a_2^2}{\triangle}(d-j+1)(k+1)\, h_{j-1,k+1,l+1}(a) \\
&+(k+1)(l-s)\frac{c_2a_2^2}{\triangle}\frac{c_2a_2^2}{c_1a_1^2}\, h_{j,k+1,l}(a) \\
&+\frac{1}{2}(r-k)[\partial_2 + 2(s-l-1) - (k+1)\frac{2c_2a_2^2}{\triangle}]\, h_{j,k,l+1}(a) \\
&= 0 \qquad (0 \le j \le d,\ 0 \le k \le r-1,\ -1 \le l \le s-1),
\end{aligned}$$

$$(H_2^- : 2)_{jkl}$$

$$\begin{aligned}
&\frac{c_1a_1^2}{\triangle}(j+1)(k-r)\, h_{j+1,k,l}(a) \\
&+(k-r)(l+1)\frac{c_1a_1^2}{\triangle}\frac{c_1a_1^2}{c_2a_2^2}\, h_{j,k,l+1}(a) \\
&-\frac{1}{2}(k+1)[\partial_1 + 2l + (r-k)\frac{2c_1a_1^2}{\triangle}]\, h_{j,k+1,l}(a) \\
&= 0 \qquad (0 \le j \le d,\ 0 \le k \le r-1,\ 0 \le l \le s),
\end{aligned}$$

$$(H_3^- : 1)_{jkl}$$

$$\begin{aligned}
&\frac{c_2a_2^2}{\triangle}(j+1)(l-s)\, h_{j+1,k,l}(a) \\
&+(k+1)(l-s)\frac{c_2a_2^2}{\triangle}\frac{c_2a_2^2}{c_1a_1^2}\, h_{j,k+1,l}(a) \\
&+\frac{1}{2}(l+1)[\partial_2 + 2k + (l-s)\frac{2c_2a_2^2}{\triangle}]\, h_{j,k,l+1}(a) \\
&= 0 \qquad (0 \le j \le d,\ 0 \le k \le r,\ 0 \le l \le s-1),
\end{aligned}$$

$(H_3^- : 2)_{jkl}$

$$
\begin{aligned}
&-\frac{c_1 a_1^2}{\triangle}(d-j+1)(l+1)\, h_{j-1,k+1,l+1}(a) \\
&+(k-r)(l+1)\frac{c_1 a_1^2}{\triangle}\frac{c_1 a_1^2}{c_2 a_2^2}\, h_{j,k,l+1}(a) \\
&+\frac{1}{2}(l-s)[\partial_1 + 2(r-k-1) + (l+1)\frac{2c_1 a_1^2}{\triangle}]\, h_{j,k+1,l}(a) \\
&= 0 \qquad (0 \le j \le d,\ -1 \le k \le r-1,\ 0 \le l \le s-1),
\end{aligned}
$$

$(HS)_{jkl}$ (91)

$$
\begin{aligned}
&[(\partial_1+\partial_2)^2 + (3r+3s+u+2)(\partial_1+\partial_2) - 8\pi(c_1 a_1^2 \partial_1 + c_2 a_2^2 \partial_2) \\
&-8\pi(c_1 a_1^2 + c_2 a_2^2)(r+s+2) \\
&+8\pi(2k-2l-r+s)(c_1 a_1^2 - c_2 a_2^2) \\
&+2(r+s+1)(r+s+u)]\, h_{j,k,l}(a) = 0. \qquad (92)
\end{aligned}
$$

PROOF. Direct computation. □

We require a lemma on analyticity of $h_{jkl}(a)$.

LEMMA 11.5. *Under the same assumptions and notations in Proposition 11.4, if $h_{jkl}(a)$, which satisfies $(H_2^- : 1)_{kl}$, $(H_2^- : 2)_{kl}$, $(H_3^- : 1)_{kl}$, and $(H_3^- : 2)$, is analytic on A, then $(c_1 a_1^2 - c_2 a_2^2)^{-\frac{d-r-s}{2}} \cdot h_{jkl}(a)$ is also analytic on A, i.e. $h_{jkl}(a)$ is exactly divisible by $(c_1 a_1^2 - c_2 a_2^2)^{\frac{d-r-s}{2}}$.*

PROOF. We introduce new variables $x = 2\pi(c_1 a_1^2 - c_2 a_2^2)$ and $y = 2\pi(c_1 a_1^2 + c_2 a_2^2)$. Then $h_{jkl}(x,y) = h_{jkl}(a_1,a_2)$ is also analytic on the variable x, (note that c_1, c_2 are positive) i.e. there is a integer $N_0 \ge 0$ such that

$$
h_{jkl}(x,y) = \sum_{n=N_0}^{\infty} H_{jkl;n}(y) x^n, \qquad (93)
$$

for suitable functions $\{H_{jkl;n}(y)\}_{n=N_0}^{\infty}$. Next we substitute (93) into equations in Proposition 11.4 and we have $N_0 = \frac{d-r-s}{2}$ or $N_0 = -\frac{d+r+s}{2} - 1$ after some calculations with method. (See Lemma 8.7 in the section of P_J-principal series.) $N_0 \ge 0$ implies that $N_0 = \frac{d-r-s}{2}$. □

Above Lemma leads us to the following.

PROPOSITION 11.6. *Suppose that $d < r+s$. If the functions $c_{jkl}(a)$ satisfy the differential equations $(C_2^- : 1)$, $(C_2^- : 2)$, $(C_3^- : 1)$ and $(C_3^- : 2)$ in Proposition 11.3 and are analytic on A, then $c_{jkl}(a)$ are identically zero.*

We assume that $d \geq r+s$ hereafter and put $c_{k,l}(a) = c_{\frac{d+r+s}{2}-k-l,k,l}(a)$ i.e. $h_{k,l}(a) = h_{\frac{d+r+s}{2}-k-l,k,l}(a)$ for simplicity.

We want to determine $h_{0,0}(a)$ (or $h_{r,s}(a)$). Then the recurrence relations $(H_2^- : 1)$, $(H_2^- : 2)$, $(H_3^- : 1)$, and $(H_3^- : 2)$ determine all $h_{kl}(a)$.

PROPOSITION 11.7. *Under the same assumptions and notations until the above, if $h_{0,0}(a)$ satisfies $(H_2^- : 1)$, $(H_2^- : 2)$ and $(H_3^- : 1)$ in Proposition 11.4, then*

$$\Big[\partial_1\partial_2 - (s+1)\frac{2c_2a_2^2}{\triangle}\partial_1 + (r+1)\frac{2c_1a_1^2}{\triangle}\partial_2 + 4c_1c_2(\frac{a_1a_2}{\triangle})^2(\frac{d-r-s}{2})(\frac{d+r+s}{2}+1)\Big]\, h_{0,0}(a) = 0. \quad (94)$$

and if $h_{r,s}(a)$ satisfies $(H_2^- : 2)$, $(H_3^- : 1)$ and $(H_3^- : 2)$ in Proposition 11.4, then

$$\Big[\partial_1\partial_2 - (r+1)\frac{2c_2a_2^2}{\triangle}\partial_1 + (s+1)\frac{2c_1a_1^2}{\triangle}\partial_2 + 4c_1c_2(\frac{a_1a_2}{\triangle})^2(\frac{d-r-s}{2})(\frac{d+r+s}{2}+1)\Big]\, h_{r,s}(a) = 0. \quad (95)$$

PROOF. We show the case of $h_{0,0}(a)$. The calculation for $h_{r,s}(a)$ is similar. Firstly rewrite $(H_2^- : 1)_{jkl}$ into $(H_2^- : 1)_{kl}$ by setting $j = \frac{d+r+s}{2} - (k+1) - (l+1) + 1$ etc. Subtract $(H_2^- : 2) \times (l-s)(c_2a_2^2)$ from $(H_3^- : 1) \times (k-r)(c_1a_1^2)$, then we get

$$\begin{aligned}(k+1)(s-l)\frac{2c_2a_2^2}{\triangle}&[\partial_1 - 2(k-l-r)]\, h_{k+1,l}(a)\\ &= (k-r)(l+1)\frac{2c_1a_1^2}{\triangle}[\partial_2 + 2(k-l+s)]\, h_{k,l+1}(a). \qquad (96)\end{aligned}$$

We have the same equation from $(H_2^- : 1)$ and $(H_3^- : 2)$. So we show (94) from three equations $(H_2^- : 1)_{l=-1}$, $(H_2^- : 2)_{l=0}$ and $(96)_{l=0}$. Operate $[\partial_2 + 2(k+s)][\frac{\triangle}{c_1a_1^2}\frac{c_2a_2^2}{c_1a_1^2}]$ on $(H_2^- : 2)_{l=0}$,

$$\begin{aligned}&(\frac{d+r+s}{2} - k)(k-r)[\partial_2 + 2(k+s)]\frac{c_2a_2^2}{c_1a_1^2}\, h_{k,0}(a)\\ &+ (k-r)[\partial_2 + 2(k+s)]\, h_{k,1}(a)\\ &- \frac{1}{2}(k+1)[\partial_2 + 2(k+s)]\\ &\qquad \times \frac{\triangle}{c_1a_1^2}\frac{c_2a_2^2}{c_1a_1^2}\Big[\partial_1 + (r-k)\frac{2c_1a_1^2}{\triangle}\Big]\, h_{k+1,0}(a) = 0. \qquad (97)\end{aligned}$$

By using $(H_2^- : 1)_{l=-1}$, the first line of (97) equals

$$\begin{aligned}
&(\frac{d+r+s}{2} - k)(k-r)\frac{c_2a_2^2}{c_1a_1^2}[\partial_2 + 2(k+s)+2]\, h_{k,0}(a) \\
&= (\frac{d+r+s}{2} - k)(k-r)\frac{c_2a_2^2}{c_1a_1^2}\Big[\partial_2 + 2s - (k+1)\frac{2c_2a_2^2}{\triangle} \\
&\qquad +(k+1)\frac{2c_1a_1^2}{\triangle}\Big]\, h_{k,0}(a) \\
&= \frac{c_2a_2^2}{c_1a_1^2}(\frac{d+r+s}{2} - k) \times (-1)\frac{2c_2a_2^2}{\triangle}(\frac{d-r-s}{2} + k + 1)(k+1)\, h_{k+1,0}(a) \\
&\quad +\frac{c_2a_2^2}{c_1a_1^2}(\frac{d+r+s}{2} - k)(k-r)(k+1)\frac{2c_1a_1^2}{\triangle}]\, h_{k,0}(a).
\end{aligned}$$

By using $(96)_{l=0}$, the second line of (97) equals

$$\frac{c_2a_2^2}{c_1a_1^2}(k+1)s[\partial_1 - 2(k-r)]\, h_{k+1,0}(a) \tag{98}$$

So (97) is rewritten as

$$\begin{aligned}
&-\frac{c_2a_2^2}{c_1a_1^2}\frac{2c_2a_2^2}{\triangle}(\frac{d+r+s}{2} - k)(\frac{d-r-s}{2} + k + 1)(k+1)\, h_{k+1,0}(a) \\
&+\frac{c_2a_2^2}{c_1a_1^2}(\frac{d+r+s}{2} - k)(k-r)(k+1)\frac{2c_1a_1^2}{\triangle}\, h_{k,0}(a) \\
&+\frac{c_2a_2^2}{c_1a_1^2}s(k+1)[\partial_1 - 2(k-r)]\, h_{k+1,0}(a) \\
&-\frac{1}{2}(k+1)[\partial_2 + 2(k+s)]\frac{\triangle}{c_1a_1^2}\frac{c_2a_2^2}{c_1a_1^2}[\partial_1 + (r-k)\frac{2c_1a_1^2}{\triangle}]\, h_{k+1,0}(a) = 0.
\end{aligned}$$

Thus we get

$$\begin{aligned}
&(k+1)\Big[\partial_1\partial_2 - (s+1)\frac{2c_2a_2^2}{\triangle}\partial_1 + (r-k)\frac{2c_1a_1^2}{\triangle}\partial_2 \\
&\qquad +2(k+1)\partial_1 + 2(k+1)(r-k)\frac{2c_1a_1^2}{\triangle} \\
&\qquad +4c_1c_2(\frac{a_1a_2}{\triangle})^2(\frac{d+r+s}{2} - k)(\frac{d-r-s}{2} + k + 1)\Big]\, h_{k+1,0}(a) \\
&-4(\frac{d+r+s}{2} - k)(k-r)(\frac{c_1a_1^2}{\triangle})^2\, h_{k,0}(a) = 0 \quad (0 \le k \le r-1). \quad (99)
\end{aligned}$$

Put $k=0$ in (99), then we have

$$\Big[\partial_1\partial_2-(s+1)\frac{2c_2a_2^2}{\triangle}\partial_1+r\frac{2c_1a_1^2}{\triangle}\partial_2+2\partial_1+2r\frac{2c_1a_1^2}{\triangle}$$
$$+4c_1c_2(\frac{a_1a_2}{\triangle})^2(\frac{d+r+s}{2})(\frac{d-r-s}{2}+1)\Big]\,h_{1,0}(a)$$
$$+4r(\frac{d+r+s}{2})(\frac{c_1a_1^2}{\triangle})^2\,h_{0,0}(a)=0. \tag{100}$$

We eliminate $h_{1,0}(a)$ in the above equation by using $(H_2^-:1)_{k=0,l=-1}$:

$$(\frac{d-r-s}{2}+1)\,h_{1,0}(a)=\frac{1}{2}r[\partial_2+2s-\frac{2c_2a_2^2}{\triangle}]\,h_{0,0}(a).$$

We note that $r\neq 0$ since the Harish-Chandra parameter Λ belongs to Ξ_{II} i.e. $r+s+2>-u>|r-s|+2$. Then we have the following partial differential equation for $h_{0,0}(a)$.

$$L\,h_{0,0}(a)=0. \tag{101}$$

Here, the third-order partial differential operator L is given by

$$L=\Big[\partial_1\partial_2-(s+1)\frac{2c_2a_2^2}{\triangle}\partial_1+r\frac{2c_1a_1^2}{\triangle}\partial_2+2\partial_1+2r\frac{2c_1a_1^2}{\triangle}$$
$$+4c_1c_2(\frac{a_1a_2}{\triangle})^2(\frac{d+r+s}{2})(\frac{d-r-s}{2}+1)\Big]$$
$$\times\frac{1}{2}\frac{\triangle}{c_2a_2^2}\Big[\partial_2+2s-\frac{2c_2a_2^2}{\triangle}\Big]$$
$$+4(\frac{d+r+s}{2})(\frac{d-r-s}{2}+1)(\frac{c_1a_1^2}{\triangle})^2. \tag{102}$$

Next we require a lemma on the differential operator L to get the desired equation (94) of $h_{0,0}(a)$.

LEMMA 11.8. *Define* L'

$$=\frac{\triangle}{2c_2a_2^2}\Big[\partial_2+2s-2\frac{2c_2a_2^2}{\triangle}\Big]\Big[\partial_1\partial_2-(s+1)\frac{2c_2a_2^2}{\triangle}\partial_1+(r+1)\frac{2c_1a_1^2}{\triangle}\partial_2$$
$$+4c_1c_2(\frac{a_1a_2}{\triangle})^2(\frac{d-r-s}{2})(\frac{d+r+s}{2}+1)\Big], \tag{103}$$

then we have

$$L=L'.$$

PROOF. Both differential operators equal $\frac{\triangle}{2c_2a_2^2}$ times

$$\partial_1\partial_2^2+(r+1)\frac{2c_1a_1^2}{\triangle}\partial_2^2+s(2-4\frac{2c_2a_2^2}{\triangle})\partial_1\partial_2$$
$$+\frac{2c_1a_1^2}{\triangle}\frac{2c_2a_2^2}{\triangle}\left\{(\frac{d+r+s}{2}+1)(\frac{d-r-s}{2})-(r+1)\right\}\partial_2$$
$$+2s(r+1)\frac{2c_1a_1^2}{\triangle}\partial_2-2s(s+1)\frac{2c_2a_2^2}{\triangle}\partial_1$$
$$+2(s+1)(\frac{2c_2a_2^2}{\triangle})^2\partial_1-(s+1)\frac{2c_1a_1^2}{\triangle}\frac{2c_2a_2^2}{\triangle}\partial_1$$
$$+2(s+1)\frac{2c_1a_1^2}{\triangle}\frac{2c_2a_2^2}{\triangle}(\frac{d+r+s}{2}+1)(\frac{d-r-s}{2}). \tag{104}$$

So we get the assertion. □

Let us complete the proof of Proposition 11.7. By using the above lemma, (101) is written as follows.

$$\left[\partial_2+2s-2\frac{2c_2a_2^2}{\triangle}\right]\left[\partial_1\partial_2-(s+1)\frac{2c_2a_2^2}{\triangle}\partial_1+(r+1)\frac{2c_1a_1^2}{\triangle}\partial_2\right.$$
$$\left.+4c_1c_2(\frac{a_1a_2}{\triangle})^2(\frac{d-r-s}{2})(\frac{d+r+s}{2}+1)\right]h_{0,0}(a)=0. \tag{105}$$

Define $\tilde{h}(a)$ by

$$\tilde{h}(a)=\left[\partial_1\partial_2-(s+1)\frac{2c_2a_2^2}{\triangle}\partial_1+(r+1)\frac{2c_1a_1^2}{\triangle}\partial_2\right.$$
$$\left.+4c_1c_2(\frac{a_1a_2}{\triangle})^2(\frac{d-r-s}{2})(\frac{d+r+s}{2}+1)\right]h_{0,0}(a). \tag{106}$$

We remark that if $h_{0,0}(a)$ is analytic then $\triangle\tilde{h}(a)$ is so. (See Lemma 11.5 when $\frac{d-r-s}{2}$ is a positive integer.)

From (105),

$$\left[\partial_2+2s-2\frac{2c_2a_2^2}{\triangle}\right]\tilde{h}(a)=0. \tag{107}$$

We can check easily that any solution of the above differential equation is given by

$$\tilde{h}(a)=C(a_1)a_2^{-2s}(c_1a_1^2-c_2a_2^2)^{-2}. \tag{108}$$

Here, $C(a_1)$ is a function of a_1. If $C(a_1)$ is not identically zero, $\tilde{h}(a)$ has a double pole at $\{c_1a_1^2-c_2a_2^2=0\}$ because c_1 and c_2 are positive.

Therefore, $C(a_1)$ must be zero, i.e. $\tilde{h}(a)$ is identically zero:

$$\tilde{h}(a) = \left[\partial_1\partial_2 - (s+1)\frac{2c_2a_2^2}{\triangle}\partial_1 + (r+1)\frac{2c_1a_1^2}{\triangle}\partial_2 \right.$$
$$\left. +4c_1c_2(\frac{a_1a_2}{\triangle})^2(\frac{d-r-s}{2})(\frac{d+r+s}{2}+1)\right] h_{0,0}(a)$$
$$= 0. \tag{109}$$

This is the desired differential equation. □

11.4. Integral representation of analytic solutions for the system of differential equations. We have two partial differential equations (92) and (94) (resp. (95)) which characterize $h_{0,0}(a)$ (resp.$h_{r,s}(a)$) in the last subsection. We want to have an explicit integral representation of $h_{0,0}(a)$ (resp. $h_{r,s}(a)$). It also gives us all $h_{kl}(a)$'s integral representation.

We introduce a more tractable function $f_{kl}(a)$ by using Lemma 11.5 i.e.

$$f_{kl}(a) = [\triangle^{\frac{d-r-s}{2}}]^{-1}h_{kl}(a).$$

Let us rewrite two partial differential equations that of $f_{0,0}(a)$, changing variables $a = (a_1, a_2)$ into $y = (y_1, y_2)$ by $y_i = 2\pi c_i a_i^2$ $(i = 1, 2)$ for simplicity. Then we have the following proposition on $f_{0,0}(y)$.

PROPOSITION 11.9. *Under the same assumptions and notations until the above, $f_{0,0}(y)$ satisfies the following two partial differential equations.*

$$\left[\frac{\partial^2}{\partial y_1\partial y_2} - (\frac{d-r-s}{2}+s+1)\frac{1}{y_1-y_2}\frac{\partial}{\partial y_1}\right.$$
$$\left. +(\frac{d-r-s}{2}+r+1)\frac{1}{y_1-y_2}\frac{\partial}{\partial y_2}\right] f_{0,0}(y) = 0, \tag{110}$$

and

$$\left[y_1^2\frac{\partial^2}{\partial y_1^2} + y_2^2\frac{\partial^2}{\partial y_2^2} + 2y_1y_2\frac{\partial^2}{\partial y_1\partial y_2} + (d+b_0+2)(y_1\frac{\partial}{\partial y_1}+y_2\frac{\partial}{\partial y_2})\right.$$
$$-2\left\{(y_1^2\frac{\partial}{\partial y_1}+y_2^2\frac{\partial}{\partial y_2}) + (\frac{d}{2}+1)(y_1+y_2) + \frac{1}{2}(r-s)(y_1-y_2)\right\}$$
$$\left. +\frac{1}{4}(d+u)(d+r+s+2)\right] f_{0,0}(y) = 0. \tag{111}$$

Remark. We also have two partial differential equations of $f_{r,s}(y)$.

PROOF. Set $\alpha = d-r-s$, then we have

$$\partial_1(\triangle^{\frac{\alpha}{2}}f(a)) = \frac{\alpha}{2}\triangle^{\frac{\alpha}{2}-1}(2c_1a_1^2)f(a) + \triangle^{\frac{\alpha}{2}}(\partial_1 f(a)),$$

$$\partial_2(\triangle^{\frac{\alpha}{2}} f(a)) = \frac{\alpha}{2}\triangle^{\frac{\alpha}{2}-1}(-2c_2a_2^2)f(a) + \triangle^{\frac{\alpha}{2}}(\partial_2 f(a)),$$
$$(\partial_1+\partial_2)(\triangle^{\frac{\alpha}{2}} f(a)) = \triangle^{\frac{\alpha}{2}}[(\partial_1+\partial_2)+\alpha]f(a),$$

$$2\partial_1\partial_2(\triangle^{\frac{\alpha}{2}} f(a))$$
$$= \triangle^{\frac{\alpha}{2}}\left[2\partial_1\partial_2 - \alpha\frac{2c_2a_2^2}{\triangle}\partial_1 + \alpha\frac{2c_1a_1^2}{\triangle}\partial_2 - 2\alpha(\alpha-2)c_1c_2(\frac{a_1a_2}{\triangle})^2\right] f(a),$$

Thus we have

$$\Big[\partial_1\partial_2 - (\frac{d-r-s}{2}+s+1)\frac{2c_2a_2^2}{\triangle}\partial_1 + (\frac{d-r-s}{2}+r+1)\frac{2c_1a_1^2}{\triangle}\partial_2$$
$$+4c_1c_2(\frac{a_1a_1}{\triangle})^2(\frac{d-r-s}{2})(\frac{d+r+s}{2}+1)$$
$$+4c_1c_2(\frac{a_1a_2}{\triangle})^2\{-\frac{d-r-s}{2}(1+s) - \frac{d-r-s}{2}(1+r)$$
$$-\frac{d-r-s}{2}(\frac{d-r-s}{2}-1)\}\Big] f_{0,0}(a) = 0.$$

The constant term of the above differential operator equals 0. We use the following formula in changing the variables,

$$\partial_i f = 2y_i\frac{\partial}{\partial y_i}f,$$

for $i = 1, 2$. □

The differential equation (110) in Proposition 11.9 is called the *Euler-Darboux* equation which appears in hypergeometric function theory.

We write (110), (111) as

$$P_A\, f_{0,0}(y) = 0, \qquad P_B\, f_{0,0}(y) = 0, \tag{112}$$

i.e. P_A, P_B are partial differential operators as is defined in Proposition 11.9. For example,

$$P_A = \frac{\partial^2}{\partial y_1\partial y_2} - \Big(\frac{d-r-s}{2}+s+1\Big)\frac{1}{y_1-y_2}\frac{\partial}{\partial y_1}$$
$$+\Big(\frac{d-r-s}{2}+r+1\Big)\frac{1}{y_1-y_2}\frac{\partial}{\partial y_2}.$$

LEMMA 11.10 (Debiard etc. [**DG1, DG2**], Iida [**Ii**, Lemma 8.4]). *For the function $f_{0,0}(y_1,y_2)$ which belongs to* $\ker P_A$, *if there is a real number r such that $0 < r < \infty$ and $f_{0,0}(y_1,y_2)$ is analytic around the point (r,r), then $f_{0,0}(y_1,y_2)$ has the following integral expression.*

$$f_{0,0}(y) = \int_0^1 F(ty_1+(1-t)y_2)t^{\frac{d-r-s}{2}+r}(1-t)^{\frac{d-r-s}{2}+s}\,dt. \tag{113}$$

Here F is some analytic function around the point r.

Our function $f_{0,0}(y_1, y_2)$ satisfies the assumption in the above lemma. $f_{0,0}(y_1, y_2)$ is analytic on $(\mathbb{R}_{>0})^2$. Let us determine the integrand F.

LEMMA 11.11. *Let P_B be the differential operator defined as (112), if*

$$f_{0,0}(y) = \int_0^1 F(ty_1 + (1-t)y_2) t^{\frac{d-r-s}{2}+r} (1-t)^{\frac{d-r-s}{2}+s}\, dt$$

and

$$(P_B\, f_{0,0})(y) = \int_0^1 G(ty_1 + (1-t)y_2) t^{\frac{d-r-s}{2}+r} (1-t)^{\frac{d-r-s}{2}+s}\, dt$$

Then we have

$$G(x) = \left[x^2 \frac{d^2}{dx^2} - \{2x - (d + b_0 + 2)\} x \frac{d}{dx} - 2(d+2)x + \frac{1}{4}\{(d + b_0 + 1)^2 - (b_3 - 1)^2\}\right] F(x). \tag{114}$$

PROOF. By interchanging integration and differentiation. □

We calculate on a parallel with the case of $f_{r,s}(y)$. We have the proposition.

PROPOSITION 11.12. *Under the same assumptions and notations in Proposition 11.9, The analytic function $f_{0,0}(y)$ and $f_{r,s}(y)$ has the following integral representation:*

$$f_{0,0}(y) = \int_0^1 F(ty_1 + (1-t)y_2) t^{\frac{d-r-s}{2}+r} (1-t)^{\frac{d-r-s}{2}+s}\, dt,$$

$$f_{r,s}(y) = \int_0^1 F(ty_1 + (1-t)y_2) t^{\frac{d-r-s}{2}+s} (1-t)^{\frac{d-r-s}{2}+r}\, dt,$$

and the common integrand $F(x)$ satisfies the following differential equation

$$\left[x^2 \frac{d^2}{dx^2} - \{2x - (d + b_0 + 2)\} x \frac{d}{dx} - 2(d+2)x + \frac{1}{4}\{(d + b_0 + 1)^2 - (b_3 - 1)^2\}\right] F(x) = 0. \tag{115}$$

Next lemma tells us explicit formula for $f_{0,0}(y)$ and $f_{r,s}(y)$.

LEMMA 11.13. *The differential equation:*

$$\left[x^2 \frac{d^2}{dx^2} - \{2x - (d + b_0 + 2)\} x \frac{d}{dx} - 2(d+2)x + \frac{1}{4}\{(d + b_0 + 1)^2 - (b_3 - 1)^2\}\right] F(x) = 0. \tag{116}$$

has two linearly independent solutions:

$$F_W(x) = e^x x^{-\frac{d+b_0+2}{2}} W_{\frac{b_0-d-2}{2}, \frac{b_3-1}{2}}(2x) \tag{117}$$

and

$$F_M(x) = e^x x^{-\frac{d+b_0+2}{2}} M_{\frac{b_0-d-2}{2}, \frac{b_3-1}{2}}(2x). \tag{118}$$

Here, $W_{\kappa,\mu}(z)$ and $M_{\kappa,\mu}(z)$ are Whittaker's confluent hypergeometric functions. (See [**WW**] *Chapter 16 for definition.)*

PROOF. Put $F(x) = e^x x^{-\frac{d+b_0+2}{2}} H(x)$. We transform the differential equation to that of $H(x)$. Then $H(x)$ satisfies:

$$\frac{d^2H(x)}{dx^2} + \left\{ -1 + \frac{b_0-d-2}{x} + \frac{\frac{1}{4} - \frac{(b_3-1)^2}{4}}{x^2} \right\} H(x) = 0.$$

If we set $t = 2x$ then,

$$\frac{d^2H}{dt^2} + \left\{ -\frac{1}{4} + \frac{\frac{b_0-d-2}{2}}{t} + \frac{\frac{1}{4} - \frac{(b_3-1)^2}{4}}{t^2} \right\} H = 0.$$

This is precise Whittaker's differential equation. □

Next we rewrite Proposition 11.4 into that of $f_{kl}(y)$ to get the recurrence relations from $f_{0,0}(y)$ to $f_{kl}(y)$.

PROPOSITION 11.14.

$(F_2^- : 1)_{kl}$

$$\begin{aligned}
&-(\frac{d-r-s}{2} + k + l + 2)(k+1)\, f_{k+1,l+1}(y) \\
&+(k+1)(l-s)y_1^{-1}y_2\, f_{k+1,l}(y) \\
&+(r-k)(s-l-1)y_1y_2^{-1}\, f_{k,l+1}(y) \\
&+(r-k)\left[(y_1-y_2)\frac{\partial}{\partial y_2} - (\frac{d-r-s}{2} + s + k - l)\right] f_{k,l+1}(y) \\
&= 0 \qquad (0 \le k \le r-1,\ -1 \le l \le s-1),
\end{aligned}$$

$(F_2^- : 2)_{kl}$

$$\begin{aligned}
&(\frac{d+r+s}{2} - k - l)(k-r)\, f_{k,l}(y) \\
&+(k-r)(l+1)y_1y_2^{-1}\, f_{k,l+1}(y) \\
&+(k+1)ly_1^{-1}y_2\, f_{k+1,l}(y) \\
&-(k+1)\left[(y_1-y_2)\frac{\partial}{\partial y_1} + (\frac{d-r-s}{2} + r - k + l)\right] f_{k+1,l}(y)
\end{aligned}$$

$$= 0 \qquad (0 \le k \le r-1,\, 0 \le l \le s),$$

$(F_3^- : 1)_{kl}$

$$\begin{aligned}
&(\frac{d+r+s}{2} - k - l)(l-s)\, f_{k,l}(y) \\
&+(k+1)(l-s) y_1^{-1} y_2\, f_{k+1,l}(y) \\
&+k(l+1) y_1 y_2^{-1}\, f_{k,l+1}(y) \\
&+(l+1)\left[(y_1 - y_2)\frac{\partial}{\partial y_2} - (\frac{d-r-s}{2} + s + k - l)\right] f_{k,l+1}(y) \\
&= 0 \qquad (0 \le k \le r,\, 0 \le l \le s-1),
\end{aligned}$$

$(F_3^- : 2)_{kl}$

$$\begin{aligned}
&-(\frac{d-r-s}{2} + k + l + 2)(l+1)\, f_{k+1,l+1}(y) \\
&+(k-r)(l+1) y_1 y_2^{-1}\, f_{k,l+1}(y) \\
&+(s-l)(r-k-1) y_1^{-1} y_2\, f_{k+1,l}(y) \\
&+(l-s)\left[(y_1 - y_2)\frac{\partial}{\partial y_1} + (\frac{d-r-s}{2} + r - k + l)\right] f_{k+1,l}(y) \\
&= 0 \qquad (-1 \le k \le r-1,\, 0 \le l \le s-1).
\end{aligned}$$

PROOF. Note that

$$\partial_1 f = a_1 \frac{\partial f}{\partial a_1} = a_1 \frac{\partial f}{\partial y_1}\frac{\partial y_1}{\partial a_1} = 4\pi c_1 a_1^2 \frac{\partial f}{\partial y_1} = 2 y_1 \frac{\partial f}{\partial y_1}. \qquad \square$$

We deduce $f_{k,0}(y)$ from $f_{0,0}(y)$.

PROPOSITION 11.15. *Under the same assumptions and notations until the above,*

$$\begin{aligned}
f_{k,0}(y) = \binom{r}{k}\binom{d-s}{\frac{d-r-s}{2}}^{-1}\binom{d}{s}^{-1} \sum_{i=0}^{k} (-1)^i \binom{k}{i} \binom{\frac{d-r-s}{2}+s+i}{\frac{d-r-s}{2}+k} \\
\times \binom{d}{\frac{d-r-s}{2}+s+i} (y_1 y_2^{-1})^{k-i}\, g_{-i,i}(y). \qquad (119)
\end{aligned}$$

Here,

$$g_{-i,i}(y) = \int_0^1 F(ty_1 + (1-t)y_2) t^{\frac{d-r-s}{2}+r-i} (1-t)^{\frac{d-r-s}{2}+s+i}\, dt \qquad (120)$$

and

$$F(x) = e^x x^{-\frac{d+b_0+2}{2}} W_{\frac{b_0-d-2}{2}, \frac{b_3-1}{2}}(2x). \qquad (121)$$

($F(x) = F_W(x)$ or $F_M(x)$.)

PROOF. We prove this by induction on k. if $k = 0$ then the assertion is proved in Proposition 11.12 and Lemma 11.13. If we use $(F_2^- : 1)_{l=-1}$ in Proposition 11.14,

$$(\frac{d-r-s}{2} + k + 1)(k+1)\, f_{k+1,0}(y) = (r-k)\left[(y_1 - y_2)\frac{\partial}{\partial y_2} - (\frac{d-r-s}{2} + s + k + 1) + s y_1 y_2^{-1}\right] f_{k,0}(y). \quad (122)$$

and

$$(y_1 - y_2)\frac{\partial}{\partial y_2}\, g_{-i,i}(y) = -(\frac{d-r-s}{2} + r - i)\, g_{-i-1,i+1}(y) + (\frac{d-r-s}{2} + s + i + 1)\, g_{-i,i}(y), \quad (123)$$

then we get the assertion for $k+1$. □

Next we determine all $f_{k,l}(y)$ by using $f_{k,0}(y)$.

PROPOSITION 11.16. *Under the same assumptions and notations until the above,*

$$f_{k,l}(y) = \binom{r}{k}\binom{d-s}{\frac{d-r-s}{2}}^{-1}\binom{d}{s}^{-1} \times \sum_{i=-l}^{k} (-1)^i a_i^{(k,l)} \binom{d}{\frac{d-r-s}{2} + s + i} (y_1 y_2^{-1})^{k-l-i}\, g_{-i,i}(y) \quad (124)$$

Here,

$$a_i^{(k,l)} = \sum_{j=0}^{l} \binom{\frac{d-r-s}{2} + r - i}{j}\binom{r-i-j}{l-j}\binom{k}{i+j}\binom{\frac{d-r-s}{2} + s + i + j}{\frac{d-r-s}{2} + k + l}, \quad (125)$$

and $g_{-i,i}(y)$ is defined in (120)

PROOF. We prove them by induction on l. if $l = 0$ then the assertion is proved in Proposition 11.15. Using $(F_2^- : 2)_{kl}$ in Proposition 11.14:

$$(r-k)(l+1) y_1 y_2^{-1}\, f_{k,l+1}(y) = (\frac{d+r+s}{2} - k - l)(k-r)\, f_{k,l}(y) + (k+1) l y_1^{-1} y_2\, f_{k+1,l}(y)$$

$$-(k+1)\left[(y_1-y_2)\frac{\partial}{\partial y_1}+(\frac{d-r-s}{2}+r-k+l)\right] f_{k,l}(y), \quad (126)$$

and

$$\begin{aligned}(y_1-y_2)\frac{\partial}{\partial y_1}\, g_{-i,i}(y) &= -(\frac{d-r-s}{2}+r-i+1)\, g_{-i,i}(y)\\ &\quad +(\frac{d-r-s}{2}+s+i)\, g_{-i+1,i-1}(y), \end{aligned} \quad (127)$$

then we get the assertion for $f_{k,l+1}(y)$ after some calculations. □

11.5. Main theorem for the large discrete series representations. We summarize the results obtained in this section.

THEOREM 11.17. *Let* $\pi = \pi^*_\Lambda$, $\Lambda = [r,s;u-2] \in \Xi_{\mathrm{II}}$, *a positive definite character* $\xi = \mathrm{diag}(c_1,c_2)$ *of* N_S *and* χ *be an irreducible finite-dimensional representation of* $SU(\xi)$ ($\simeq SU(2)$) *with* $\dim\chi = d+1$. *Set* $\eta = \chi\otimes\xi$.
If $d < \frac{d+r+s}{2}$ *then*

$$\dim_{\mathbb{C}} \mathrm{Hom}_{(\mathfrak{g},K)}(\pi^*_\Lambda, C^\infty_\eta(R\backslash G)) = 0,$$

and if $d \geq \frac{d+r+s}{2}$ *and* $d \equiv r+s \pmod 2$ *then*

$$\dim_{\mathbb{C}} \mathrm{Hom}_{(\mathfrak{g},K)}(\pi^*_\Lambda, C^\infty_\eta(R\backslash G)) = 2.$$

We have moreover,

THEOREM 11.18. *We assume that* $\frac{d+r+s}{2} \in \mathbb{Z}$, $d \geq r+s$ *and* $\xi = \mathrm{diag}(c_1,c_2) > 0$. *Then,*

$$\dim_{\mathbb{C}} \mathrm{Hom}_{(\mathfrak{g},K)}(\pi^*_\Lambda, C^\infty_\eta(R\backslash G)^{\mathrm{mod}}) = 1.$$

More precisely, the generalized Whittaker functions $\Phi_{\pi,\tau}(a)$ *for the representation* $\pi = \pi^*_\Lambda$, $\Lambda = [r,s;u-2] \in \Xi_{II}$, *with minimal* K*-type* τ^*, $\tau = [r,s;u]$ *are described as follows.*
Let $\Phi_{\pi,\tau}(a) = \sum_{\substack{0\leq k\leq r\\ 0\leq l\leq s}} c_{kl}(a)(v_{\frac{d+r+s}{2}-k-l}\otimes f_{kl}) \in C^\infty_{M_0}(A; V_\chi\otimes W_\tau)$, *if* $\Phi_{\pi,\tau}(a)$ *is slowly increasing when each* a_1, a_2 *tends to infinity, then* $\Phi_{\pi,\tau}(a)$ *are up to constant uniquely determined and the coefficient functions are given as*

$$\begin{aligned} c_{kl}(a) =& \binom{r}{k}\binom{d-s}{\frac{d-r-s}{2}}^{-1}\binom{d}{s}^{-1} (\sqrt{c_1}a_1)^{b_0+r-k+l+2}(\sqrt{c_2}a_2)^{b_0+k+s-l+2}\\ &\times (c_1a_1^2-c_2a_2^2)^{\frac{d-r-s}{2}} e^{-2\pi(c_1a_1^2+c_2a_2^2)} f_{kl}(a), \end{aligned}$$

$$f_{kl}(a) = \sum_{i=-l}^{k} (-1)^i a_i^{(k,l)} (c_1 a_1^2 c_2^{-1} a_2^{-2})^{k-l-i} \binom{d}{\frac{d-r-s}{2}+s+i} g_{-i,i}(a),$$

$$a_i^{(k,l)} = \sum_{j=0}^{l} \binom{\frac{d-r-s}{2}+r-i}{j} \binom{r-i-j}{l-j} \binom{k}{i+j} \binom{\frac{d-r-s}{2}+s+i+j}{\frac{d-r-s}{2}+k+l},$$

$$g_{-i,i}(a) = \int_0^1 F(2\pi c_1 a_1^2 t + 2\pi c_2 a_2^2 (1-t)) t^{\frac{d-r-s}{2}+r-i} (1-t)^{\frac{d-r-s}{2}+s+i} \, dt.$$

Here,

$$F(x) = e^x x^{-\frac{d+b_0+2}{2}} W_{\frac{b_0-d-2}{2}, \frac{b_3-1}{2}}(2x).$$

And $W_{\kappa,\mu}(z)$ is Whittaker's confluent hypergeometric function. The integers $b_0, \ldots, b_3$ are defined in (34).

PROOF. The above formula of $c_{kl}(a)$ is obtained from $c_{00}(a)$ by using the recurrence relations. So we should check that this formula is also true for $c_{rs}(a)$. We calculate $a_i^{(r,s)}$ for $i = -s, \ldots, r$ namely

$$a_i^{(r,s)} = \sum_{j=0}^{s} \binom{\frac{d-r-s}{2}+r-i}{j} \binom{r-i-j}{s-j} \binom{r}{i+j} \binom{\frac{d-r-s}{2}+s+i+j}{\frac{d-r-s}{2}+r+s}. \tag{128}$$

Since the indices of nonvanishing summand must satisfy that

$$r-i-j \geq s-j, \quad r \geq i+j, \quad \text{and} \quad s+i+j \geq r+s.$$

We have $j = 0$ and $i = r - s$. Thus, as a result

$$a_i^{(r,s)} = \begin{cases} \binom{\frac{d-r-s}{2}+s}{s} & i = r-s. \\ 0 & i = -s, \ldots, r \text{ except for } \quad i = r-s. \end{cases} \tag{129}$$

Then

$$\begin{aligned} f_{rs}(a_1, a_2) &= (-1)^{r-s} a_{r-s}^{(r,s)} \binom{d}{\frac{d-r-s}{2}+r} g_{-r+s,r-s}(a) \\ &= C_{r,s} \int_0^1 F(2\pi c_1 a_1^2 t + 2\pi c_2 a_2^2 (1-t)) t^{\frac{d-r-s}{2}+s} (1-t)^{\frac{d-r-s}{2}+r} \, dt. \end{aligned}$$

Here, $C_{r,s}$ is a constant. So we complete the proof. □

We have the following formula by direct computation.

COROLLARY 11.19. *If we put $x = c_1 a_1^2 = c_2 a_2^2$ in $c_{kl}(a)$ in the above formula when $d = r + s$, then we have*

$$c_{kl}(x) = (-1)^{k+l} \binom{r}{k} \binom{s}{l} x^{\frac{b_0+2}{2}} e^{-2\pi x} W_{\frac{b_3-2}{2}, \frac{b_3-1}{2}}(4\pi x), \tag{130}$$

up to a constant multiple independent of k and l.

12. Generalized Whittaker functions for the middle discrete representations

12.1. Middle discrete series representations.

PROPOSITION 12.1. *Generalized Whittaker functions $\Phi_{\pi,\tau}$ for $\pi = \pi^*_\Lambda$, $\Lambda \in \Xi_{\mathrm{III}}$, with their minimal K-type τ^*, $\tau = [r, s; u]$ satisfy the following equations.*

$$P^{(\mathrm{III})}\nabla\Phi_{\pi,\tau}(a) = 0. \tag{131}$$

Here, $P^{(\mathrm{III})} = P^{(-,-)} \oplus P^{(-,+)} \oplus \bar{P}^{(-,-)} \oplus \bar{P}^{(-,+)}$.

PROOF.

$$\Delta^+_{\mathrm{III}} = \{[1,-1;2],[1,1;2],[2,0;0],[0,2;0],[1,1;-2],[1,-1;-2]\}$$

and that

$$\Delta^+_{\mathrm{III},n} = \{[1,-1;2],[1,1:2],[1,1;-2],[1,-1;-2]\}.$$

□

13. Generalized Whittaker functions for the middle discrete series associated with definite H_ξ

13.1. Explicit differential equations. We write explicit differential equations of generalized Whittaker functions $\Phi_{\pi,\tau}(a)$ for the case: $\lceil \quad \pi = \pi^*_\Lambda$, $\Lambda \in \Xi_{\mathrm{III}}$ and with their minimal K-type τ^*, $\tau = [r, s; u] \quad \rfloor$.

Let $\Phi_{\pi,\tau}(a) = \sum_{j=0}^{d} \sum_{\substack{0\le k\le r \\ 0\le l\le s}} c_{jkl}(a)(v_j \otimes f_{kl}) \in C^\infty_{M_0}(A, V_\chi \otimes W_\tau)$. Here, $\eta = \chi \otimes \xi$, χ is an irreducible finite dimensional representation of $SU(\xi) \simeq SU(2)$ and ξ is a non-trivial unitary character of N_S determined by a positive definite matrix $H_\xi = \begin{pmatrix} c_1 & \\ & c_2 \end{pmatrix}$. We assume that $\dim\chi = d+1$, $\frac{d+r+s}{2} \in \mathbb{Z}$.

Then the unknown functions $c_{jkl}(a)$ satisfy the following difference-differential equations.

PROPOSITION 13.1. *The coefficient functions $c_{jkl}(a)$ in $\Phi_{\pi,\tau}(a)$ satisfy the following system of partial differential equations. Here, we put $\partial_i = a_i\frac{\partial}{\partial a_i}$: the Euler operator $(i = 1, 2)$ and $\triangle = (c_1a_1^2 - c_2a_2^2)$.*

$$\begin{aligned}
&(P^{(-,-)})_{jkl} \\
&\quad -\sqrt{c_1c_2}\frac{a_1a_2}{\triangle}(d-j+1)(k+1)(l+1)\, c_{j-1,k+1,l+1}(a) \\
&\quad +\sqrt{c_1c_2}\frac{a_1a_2}{\triangle}(j+1)(k-r)(s-l)\, c_{j+1,k,l}(a)
\end{aligned}$$

$$
\begin{aligned}
&+\frac{1}{2}(k-r)(l+1)\Big[\partial_1-4\pi c_1a_1^2+(b_1-k+l-2s-3)\\
&\qquad\qquad +(s-l+k+1)\frac{2c_1a_1^2}{\triangle}\Big]\;c_{j,k,l+1}(a)\\
&+\frac{1}{2}(k+1)(s-l)\Big[\partial_2-4\pi c_2a_2^2+(b_2+k-2r-l-3)\\
&\qquad\qquad -(r-k+l+1)\frac{2c_2a_2^2}{\triangle}\Big]\;c_{j,k+1,l}(a)\\
&=0 \qquad (0\le j\le d,\ 0\le k\le r-1,\ 0\le l\le s-1),
\end{aligned}
$$

$$
\begin{aligned}
&(P^{(-,+)})_{jkl}\\
&\quad\sqrt{c_1c_2}\frac{a_1a_2}{\triangle}(d-j+1)(k+1)\;c_{j-1,k+1,l}(a)\\
&\quad+\sqrt{c_1c_2}\frac{a_1a_2}{\triangle}(j+1)(k-r)\;c_{j+1,k,l-1}(a)\\
&\quad+\frac{1}{2}(r-k)\Big[\partial_1-4\pi c_1a_1^2+(b_1-k+l-2)\\
&\qquad\qquad +(1+k-l)\frac{2c_1a_1^2}{\triangle}\Big]\;c_{j,k,l}(a)\\
&\quad+\frac{1}{2}(k+1)\Big[\partial_2-4\pi c_2a_2^2+(b_2+k-l-2r+2s)\\
&\qquad\qquad +(1+k-l-r+s)\frac{2c_2a_2^2}{\triangle}\Big]\;c_{j,k+1,l-1}(a)\\
&\quad=0 \qquad (0\le j\le d,\ 0\le k\le r-1,\ 0\le l\le s+1),
\end{aligned}
$$

$$
\begin{aligned}
&(\bar{P}^{(-,-)})_{jkl}\\
&\quad-\sqrt{c_1c_2}\frac{a_1a_2}{\triangle}(d-j+1)(k+1)(l+1)\;c_{j-1,k+1,l+1}(a)\\
&\quad+\sqrt{c_1c_2}\frac{a_1a_2}{\triangle}(j+1)(k-r)(s-l)\;c_{j+1,k,l}(a)\\
&\quad+\frac{1}{2}(k+1)(l-s)\Big[\partial_1+4\pi c_1a_1^2-(b_1-k+l+2r+3)\\
&\qquad\qquad +(r-k+l+1)\frac{2c_1a_1^2}{\triangle}\Big]\;c_{j,k+1,l}(a)\\
&\quad+\frac{1}{2}(r-k)(l+1)\Big[\partial_2+4\pi c_2a_2^2-(b_2+k-l+2s+3)\\
&\qquad\qquad -(k+s-l+1)\frac{2c_2a_2^2}{\triangle}\Big]\;c_{j,k,l+1}(a)
\end{aligned}
$$

$$= 0 \qquad (0 \le j \le d,\, 0 \le k \le r-1,\, 0 \le l \le s-1),$$

$$\begin{aligned}
&(\bar{P}^{(-,+)})_{jkl} \\
&\quad \sqrt{c_1c_2}\frac{a_1a_2}{\triangle}(d-j+1)(k+1)\, c_{j-1,k+1,l}(a) \\
&\quad +\sqrt{c_1c_2}\frac{a_1a_2}{\triangle}(j+1)(k-r)\, c_{j+1,k,l-1}(a) \\
&\quad -\frac{1}{2}(k+1)\Big[\partial_1 + 4\pi c_1 a_1^2 - (b_1 - k + l + 2r - 2s) \\
&\qquad\qquad +(r-k-s+l-1)\frac{2c_1a_1^2}{\triangle}\Big]\, c_{j,k+1,l-1}(a) \\
&\quad +\frac{1}{2}(k-r)\Big[\partial_2 + 4\pi c_2 a_2^2 - (b_2 + k - l + 2) \\
&\qquad\qquad -(k-l+1)\frac{2c_2a_2^2}{\triangle}\Big]\, c_{j,k,l}(a) \\
&\quad = 0 \qquad (0 \le j \le d,\, 0 \le k \le r-1,\, 0 \le l \le s+1).
\end{aligned}$$

PROOF. Proposition 12.1 says $P \circ \mathrm{Rad}(\nabla)\Phi_{\pi,\tau}(a) = 0$, for $P = P^{(-,-)}$, $P^{(-,+)}$, $\bar{P}^{(-,-)}$ and $\bar{P}^{(-,+)}$. For example, we write down concretely the case $P = P^{(-,+)}$ along Lemma 7.2.

$$\begin{aligned}
&P^{(-,+)} \circ \mathrm{Rad}(\nabla^+)\Phi(a) \\
&\quad = \sum_{j,k,l} P^{(-,+)} \circ \mathrm{Rad}(\nabla^+) c_{jkl}(a)(v_j \otimes f_{kl}) \\
&\quad = \sum_{j,k,l} [A^{(-,+)}_{jkl}(a)(v_{j+1} \otimes f_{k-1,l}) + B^{(-,+)}_{jkl}(a)(v_{j-1} \otimes f_{k,l+1}) \\
&\qquad\qquad + C^{(-,+)}_{jkl}(a)(v_j \otimes f_{k,l}) + D^{(-,+)}_{jkl}(a)(v_j \otimes f_{k-1,l+1})],
\end{aligned}$$

So $P^{(-,+)} \circ \mathrm{Rad}(\nabla^+)\Phi(a) = 0$ is equivalent to

$$A^{(-,+)}_{j-1,k+1,l}(a) + B^{(-,+)}_{j+1,k,l-1}(a) + C^{(-,+)}_{j,k,l}(a) + D^{(-,+)}_{j,k+1,l-1}(a) = 0. \qquad (132)$$

This is precise $(P^{(-,+)})_{jkl}$. Other three difference differential equations of $c_{jkl}(a)$ are obtained with similar manner. □

We firstly reduce the system to the following.

PROPOSITION 13.2. *The system in Proposition 13.1 is equivalent to the following system of partial differential equations.*

$$\begin{aligned}
&(C_2^+ : 1)_{jkl} \\
&\qquad -\sqrt{c_1c_2}\frac{a_1a_2}{\triangle}(d-j+1)(k+1)\, c_{j-1,k+1,l+1}(a)
\end{aligned}$$

$$+(k+1)(l-s)\frac{c_1a_1^2}{\triangle}\, c_{j,k+1,l}(a)$$

$$+\frac{1}{2}(k-r)\Big[\partial_1 - 4\pi c_1 a_1^2 + (b_1-k-l-3)$$

$$+(k+1)\frac{2c_1a_1^2}{\triangle}\Big]\, c_{j,k,l+1}(a)$$

$$=0 \qquad (0\le j\le d,\ 0\le k\le r-1,\ -1\le l\le s-1),$$

$(C_2^+ : 2)_{jkl}$

$$\sqrt{c_1c_2}\frac{a_1a_2}{\triangle}(j+1)(k-r)\, c_{j+1,k,l}(a)$$

$$+(k-r)(l+1)\frac{c_2a_2^2}{\triangle}\, c_{j,k,l+1}(a)$$

$$+\frac{1}{2}(k+1)\Big[\partial_2 - 4\pi c_2 a_2^2 + (b_2+k+l-2r-1)$$

$$-(r-k)\frac{2c_2a_2^2}{\triangle}\Big]\, c_{j,k+1,l}(a)$$

$$=0 \qquad (0\le j\le d,\ 0\le k\le r-1,\ 0\le l\le s),$$

$(C_2^- : 1)_{jkl}$

$$-\sqrt{c_1c_2}\frac{a_1a_2}{\triangle}(d-j+1)(k+1)\, c_{j-1,k+1,l+1}(a)$$

$$+(k+1)(l-s)\frac{c_2a_2^2}{\triangle}\, c_{j,k+1,l}(a)$$

$$+\frac{1}{2}(r-k)\Big[\partial_2 + 4\pi c_2 a_2^2 - (b_2+k+l+3)$$

$$-(k+1)\frac{2c_2a_2^2}{\triangle}\Big]\, c_{j,k,l+1}(a)$$

$$=0 \qquad (0\le j\le d,\ 0\le k\le r-1,\ -1\le l\le s-1),$$

$(C_2^- : 2)_{jkl}$

$$\sqrt{c_1c_2}\frac{a_1a_2}{\triangle}(j+1)(k-r)\, c_{j+1,k,l}(a)$$

$$+(k-r)(l+1)\frac{c_1a_1^2}{\triangle}\, c_{j,k,l+1}(a)$$

$$-\frac{1}{2}(k+1)\Big[\partial_1 + 4\pi c_1 a_1^2 - (b_1-k-l+2r+1)$$

$$+(r-k)\frac{2c_1a_1^2}{\triangle}\Big]\, c_{j,k+1,l}(a)$$
$$=0 \qquad (0\le j\le d,\ 0\le k\le r-1,\ 0\le l\le s).$$

PROOF. The equation $(C_2^- : 1)_{jkl}$ is obtained from:

$$(\bar{P}^{(-,-)})_{j,k,l} - (s-l)(\bar{P}^{(-,+)})_{j,k,l+1}.$$

The equation $(C_2^- : 2)_{jkl}$ is obtained from:

$$(\bar{P}^{(-,-)})_{j,k,l} + (l+1)(\bar{P}^{(-,+)})_{j,k,l+1}.$$

The equations $(C_2^+ : 1)$ and $(C_2^+ : 2)$ are obtained from $(P^{-,-})_{j,k,l}$ and $(P^{-,+})_{j,k,l+1}$ similarly. □

13.2. The case of holonomic system of rank 0. We study the system of partial differential equations obtained in the last subsection. Firstly we treat the case of the system having no solution other than zero.

PROPOSITION 13.3. *If $d\ge r+s$ then the coefficient functions $c_{jkl}(a)$ $(0\le j\le d,\ 0\le k\le r,\ 0\le l\le s$ and $2j+2k+2l=d+r+s)$, which satisfy the partial differential equations $(C_2^+ : 1)_{jkl},\dots,(C_2^- : 2)_{jkl}$ in Proposition 13.2, are identically zero.*

The assumption $d\ge r+s$ implies that j which satisfies $j+k+l=\frac{d+r+s}{2}$ always exist for all $0\le k\le r$, $0\le l\le s$. So we write $c_{kl}(a)=c_{\frac{d+r+s}{2}-k-l,k,l}(a)$ for simplicity hereafter in this subsection. First of all we treat the case $s=0$.

Proof of the case $s=0$.
We get from $(C_2^+ : 1)_{l=-1}-(C_2^- : 1)_{l=-1}$ and $(C_2^- : 2)_{l=0}-(C_2^+ : 1)_{l=0}$,

$$(r-k)[\partial_1+\partial_2-4\pi(c_1a_1^2-c_2a_2^2)-(r+2)]\, c_{k,0}(a)=0,$$
$$(k+1)[\partial_1+\partial_2+4\pi(c_1a_1^2-c_2a_2^2)-(r+2)]\, c_{k+1,0}(a)=0.$$

So we have for $k=1,\dots,r-1$,

$$4\pi(c_1a_1^2-c_2a_2^2)\, c_{k,0}(a)=0. \tag{133}$$

The equation $(C_2^+ : 1)$ which put $k=r-1$, $l=-1$ and (133) implies $c_{r,0}(a)=0$. We have $c_{0,0}(a)=0$ from $(C_2^+ : 2)_{k=l=0}$ similarly. □

We introduce two polynomials which are useful to prove Proposition 13.3 when $s\ne 0$.

DEFINITION 13.4 ($\mathcal{F}_q^{\pm}(x)$). For an integer $q = 0, \ldots, s$, define

$$\mathcal{F}_q^{+}(x) = \sum_{i=0}^{q} \binom{q}{i} \frac{(s-q+i)!}{(s-q)!} x^{q-i}, \tag{134}$$

$$\mathcal{F}_q^{-}(x) = \sum_{i=0}^{q} (-1)^i \binom{q}{i} \frac{(s-q+i)!}{(s-q)!} x^{q-i}. \tag{135}$$

For example,

$$\mathcal{F}_0^{\pm}(x) = 1,$$
$$\mathcal{F}_1^{\pm}(x) = x \pm s,$$
$$\mathcal{F}_2^{\pm}(x) = x^2 \pm 2(s-1)x + s(s-1),$$
$$\mathcal{F}_3^{\pm}(x) = x^3 \pm 3(s-2)x^2 + 3(s-1)(s-2)x \pm s(s-1)(s-2).$$

We note the following fact concerning on the index q of $\mathcal{F}_q^{\pm}(x)$.

LEMMA 13.5.

$$\left[-x\frac{d}{dx} + x + (s-q)\right] \mathcal{F}_q^{+}(x) = \mathcal{F}_{q+1}^{+}(x), \tag{136}$$

$$\left[x\frac{d}{dx} + x - (s-q)\right] \mathcal{F}_q^{-}(x) = \mathcal{F}_{q+1}^{-}(x). \tag{137}$$

PROOF. We show (137). By definition,

$$\begin{aligned}
&\left[x\frac{d}{dx} + x - (s-q)\right] \mathcal{F}_q^{-}(x) \\
&= \sum_{i=0}^{q} \Big[(-1)^i \binom{q}{i} \frac{(s-q+i)!}{(s-q)!} (q-i) x^{q-i} \\
&\quad + (-1)^i \binom{q}{i} \frac{(s-q+i)!}{(s-q)!} x^{q-i+1} - (-1)^i \binom{q}{i} \frac{(s-q+i)!}{(s-q)!} (s-q) x^{q-i}\Big].
\end{aligned}$$

And use the equality:

$$\begin{aligned}
&(-1)^i \binom{q}{i} \frac{(s-q+i)!}{(s-q)!} (q-i) + (-1)^{i+1} \binom{q}{i+1} \frac{(s-q+i+1)!}{(s-q)!} \\
&\quad - (-1)^i \binom{q}{i} \frac{(s-q+i)!}{(s-q)!} (s-q)! \\
&= (-1)^{i+1} \binom{q}{i+1} \frac{(s-q+i)!}{(s-q)!} \{-(i+1) + (s-q+i+1)\} \\
&\quad + (-1)^{i+1} \binom{q}{i} \frac{(s-q+i)!}{(s-q)!} (s-q)
\end{aligned}$$

$$= (-1)^{i+1}\binom{q+1}{i+1}\frac{(s-(q+1)+(i+1))!}{(s-(q+1))!}.$$

□

The proof of Proposition 13.3 when $s \neq 0$ is reduced to show the following lemma.

LEMMA 13.6. *Suppose that $s \neq 0$. We have the following formulas on coefficient functions $c_{kl}(a)$ for $0 \leq k \leq r-s$ and $q = 0, \ldots, s$.*

$$\binom{k+q}{q} q!\, c_{k+q,s-q}(a) = \binom{r-k}{q} \mathcal{F}_q^+(4\pi\triangle) c_{k,s}(a), \tag{138}$$

$$(-1)^q \binom{r-k}{q} q!\, c_{k,q}(a) = \binom{k+q}{q} \mathcal{F}_q^-(4\pi\triangle) c_{k+q,0}(a), \tag{139}$$

where $\mathcal{F}_q^\pm(x)$ are defined in Definition 13.4 and $\triangle = c_1 a_1^2 - c_2 a_2^2$.

PROOF. From $(C_2^+ : 1) - (C_2^- : 1)$ and $(C_2^- : 2) - (C_2^+ : 1)$, we calculate them as follows,

$(A)_{kl}$

$$2(k+1)(l-s)\, c_{k+1,l}(a)$$
$$= (r-k)[\partial_1 + \partial_2 - 4\pi(c_1a_1^2 - c_2a_2^2) - (r-s+2l+4)]\, c_{k,l+1}(a)$$

and

$(B)_{kl}$

$$2(k-r)(l+1)\, c_{k,l+1}(a)$$
$$= (k+1)[\partial_1 + \partial_2 + 4\pi(c_1a_1^2 - c_2a_2^2) - (r+s-2l+2)]\, c_{k+1,l}(a).$$

We show (139) by induction on q, (138) is quite similar. For $q = 0$, (139) is trivial. Let $q > 0$, we use $(B)_{l=q}$,

$$-2(r-k)(q+1)\, c_{k,q+1}(a)$$
$$= (k+1)[\partial_1 + \partial_2 + 4\pi\triangle - (r+s-2q+2)]\, c_{k+1,q}(a). \tag{140}$$

The right hand side of $(B)_{l=q}$ becomes

$$(k+1)\cdot(-1)^{-q}\binom{r-k-1}{q}^{-1}(q!)^{-1}\binom{k+1+q}{q}$$
$$\times [\partial_1 + \partial_2 + 4\pi\triangle - (r+s-2q+2)]\mathcal{F}_q^-(4\pi\triangle)\, c_{k+1+q,0}(a). \tag{141}$$

Therefore we have,

$$2(-1)^{q+1}\binom{r-k}{q+1}(q+1)!\binom{k+q+1}{q+1}^{-1} c_{k,q+1}(a)$$
$$= [\partial_1 + \partial_2 + 4\pi\triangle - (r+s-2q+2)]\mathcal{F}_q^-(4\pi\triangle)\, c_{k+q+1,0}(a)$$

$$= \{(\partial_1 + \partial_2)\mathcal{F}_q^-(4\pi\triangle)\} \cdot c_{k+q+1,0}(a)$$
$$+\mathcal{F}_q^-(4\pi\triangle)[\partial_1 + \partial_2 - 4\pi\triangle - (r - s + 2)$$
$$+8\pi\triangle + 2(q - s)]\, c_{k+q+1,0}(a)$$
$$= \{[(\partial_1 + \partial_2) + 8\pi\triangle - 2(s - q)]\, \mathcal{F}_q^-(4\pi\triangle)\} \cdot c_{k+q+1,0}(a).$$

Here, we used the fact from $(A)_{l=-1}$ and $k + q < r - 1$,

$$(r - k - q - 1)[\partial_1 + \partial_2 - 4\pi\triangle - (r - s + 2)]\, c_{k+q+1,0}(a) = 0.$$

Since $(\partial_1 + \partial_2)\, 4\pi\triangle = 8\pi\triangle$, we have only to check that:

$$\left[2x\frac{d}{dx} + 2x - 2(s - q)\right] \mathcal{F}_q^-(x) = 2\mathcal{F}_{q+1}^-(x).$$

This formula is (137) in Lemma 13.5. □

In particular, if we put $q = s$ in the above lemma, then we have the following.

$$\binom{k+s}{s} c_{k+s,0}(a) = \binom{r-k}{s} [\sum_{i=0}^{s} \frac{(4\pi(c_1 a_1^2 - c_2 a_2^2))^i}{i!}]\, c_{k,s}(a), \quad (142)$$

$$\binom{r-k}{s} c_{k,s}(a) = \binom{k+s}{s} [\sum_{i=0}^{s} \frac{(-4\pi(c_1 a_1^2 - c_2 a_2^2))^i}{i!}]\, c_{k+s,0}(a). (143)$$

We shall complete the Proposition 13.3 by using (142) and (143).

Proof of Proposition 13.3 when $s \neq 0$. We will show $c_{kl}(a) \equiv 0$ in this case.

The above equations leads to the following:

$$c_{s,0}(a) = c_{s+1,0}(a) = \cdots = c_{r,0}(a) \equiv 0, \quad (144)$$
$$c_{0,s}(a) = c_{1,s}(a) = \cdots = c_{r-s,s}(a) \equiv 0. \quad (145)$$

Furthermore (144), (145), $(A)_{kl}$ and $(B)_{kl}$ imply that

$$c_{s,0}(a) = c_{s+1,0}(a) = \cdots = c_{r,0}(a) \equiv 0.$$
$$c_{s-1,1}(a) = c_{s,1}(a) = \cdots = c_{r-1,1}(a) \equiv 0.$$
$$\vdots$$
$$c_{1,s-1}(a) = c_{2,s-1}(a) = \cdots = c_{r-s+1,s-1}(a) \equiv 0.$$
$$c_{0,s}(a) = c_{1,s}(a) = \cdots = c_{r-s,s}(a) \equiv 0.$$

We remark that $r-s-2 > |u|$, i.e. $r-s > 2$, since the Harish-Chandra parameter Λ is in Ξ_{III}.

Next $c_{s,0}(a) = c_{s-1,1}(a) = 0$ and $(C_2^+ : 2)_{k=s-1,l=0}$ implies that $c_{s-1,0}(a) = 0$. similarly we have,

$$c_{s-1,0}(a) = c_{s-2,1}(a) = \cdots = c_{0,s-1}(a) \equiv 0.$$

On the other hand $c_{r,0}(a) = c_{r-1,1}(a) = 0$ and $(C_2^+ : 1)_{k=r-1,l=0}$ implies that $c_{r,1}(a) = 0$ and that

$$c_{r,1}(a) = c_{r-1,2}(a) = \cdots = c_{r-s+1,s}(a) \equiv 0.$$

In conclusion we have

$$c_{kl}(a) \equiv 0 \quad (0 \leq k \leq r,\ 0 \leq l \leq s)$$

by iterating these steps. □

In the proof of Proposition 13.3, the key was to show two equalities (144) and (145):

$$c_{s,0}(a) = c_{s+1,0}(a) = \cdots = c_{r,0}(a) \equiv 0,$$
$$c_{0,s}(a) = c_{1,s}(a) = \cdots = c_{r-s,s}(a) \equiv 0.$$

But there can be a d in $0 \leq d < r+s$ which implies (144) and (145). Actually, if there exists a j $(0 \leq j \leq d)$ for any pair (k,l) which appears in (144) and (145) as $c_{k,l}(a) = c_{\frac{d+r+s}{2}-k-l,k,l}(a)$, i.e.

$$0 \leq \frac{d+r+s}{2} - s, \frac{d+r+s}{2} - (s+1), \ldots, \frac{d+r+s}{2} - r \leq d. \quad (146)$$

then the equations (144) and (145) holds. Concludingly,

$$d \geq r - s \qquad \text{and} \qquad d \geq -r + s.$$

implies that $c_{jkl}(a)$ vanish. Since $\Lambda \in \Xi_{\mathrm{III}}$ implies that $r - s - 2 > |u|$, $d \geq -r + s$ is trivial. Thus we have a refinement of Proposition 13.3.

PROPOSITION 13.7. *If $d \geq r - s$ then the coefficient functions $c_{jkl}(a)$ $(0 \leq j \leq d,\ 0 \leq k \leq r,\ 0 \leq l \leq s$ and $2j + 2k + 2l = d + r + s)$, which satisfy the partial differential equations $(C_2^+ : 1)_{jkl}, \ldots, (C_2^- : 2)_{jkl}$ in Proposition 13.2, are identically zero.*

13.3. The case of holonomic system of at most rank 2. We have already treated the case of $d \geq r - s$ in the last subsection. (See Proposition 13.7.) Secondly we assume that $0 \leq d < r - s$, i.e. $r > d + s$ in this subsection. This assumption means that there is a k such that $j + k + l = \frac{d+r+s}{2}$ for any pair (j,l) in $0 \leq j \leq d$ and $0 \leq l \leq s$. We remarked that k cannot be 0 or r in this case.

PROPOSITION 13.8. *Under the same assumptions and notations in Proposition 13.2, the functions $c_{jkl}(a)$ satisfy the following system of partial differential equations.*

$(CJ^-)_{jkl}$

$$4\sqrt{c_1 c_2}\, a_1 a_2 (d - j + 1)(k + 1)\, c_{j-1,k+1,l}(a)$$
$$= (r - k)\Big[2c_2 a_2^2 \partial_1 - 2c_2 a_2^2 (r + 2 - j + \frac{d-u}{2})$$

$$+2c_1a_1^2\partial_2 - 2c_1a_1^2(r+2-j+\frac{d+u}{2})\Big]\, c_{j,k,l}(a)$$
$$(0 \le j \le d,\ 0 \le l \le s,\ k = \frac{d+r+s}{2} - j - l),$$

$(CJ^+)_{jkl}$

$$\begin{aligned}
&4\sqrt{c_1c_2}a_1a_2(j+1)(k-r-1)\, c_{j+1,k-1,l}(a)\\
&= -k\Big[2c_2a_2^2\partial_1 - 2c_2a_2^2(r+2+j-\frac{d-u}{2})\\
&\qquad +2c_1a_1^2\partial_2 - 2c_1a_1^2(r+2+j-\frac{d+u}{2})\Big]\, c_{j,k,l}(a)
\end{aligned}$$
$$(0 \le j \le d,\ 0 \le l \le s,\ k = \frac{d+r+s}{2} - j - l),$$

$(CK^+L^-)_{jkl}$

$$\begin{aligned}
&2(k+1)(l-s)\, c_{j,k+1,l}(a)\\
&= (r-k)\Big[\partial_1 + \partial_2 - 4\pi(c_1a_1^2 - c_2a_2^2)\\
&\qquad -(r-s+2l+4)\Big]\, c_{j,k,l+1}(a)
\end{aligned}$$
$$(0 \le j \le d,\ -1 \le l \le s-1,\ k = \frac{d+r+s}{2} - j - l - 1),$$

$(CK^-L^+)_{jkl}$

$$\begin{aligned}
&2(k-r)(l+1)\, c_{j,k,l+1}(a)\\
&= (k+1)\Big[\partial_1 + \partial_2 + 4\pi(c_1a_1^2 - c_2a_2^2)\\
&\qquad -(r+s-2l+2)\Big]\, c_{j,k+1,l}(a)
\end{aligned}$$
$$(0 \le j \le d,\ 0 \le l \le s,\ k = \frac{d+r+s}{2} - j - l - 1),$$

$(CS)_{jkl}$

$$\begin{aligned}
&[(\partial_1+\partial_2)^2 - 2(r+3)(\partial_1+\partial_2) + 8\pi(s-2l)(c_1a_1^2 - c_2a_2^2)\\
&-16\pi^2(c_1a_1^2 - c_2a_2^2)^2 + (r+3)^2 - (s+1)^2]\, c_{j,k,l}(a) = 0.
\end{aligned}$$

Remark. $(CK^+L^-)_{jkl}$ and $(CK^-L^+)_{jkl}$ implies $(CS)_{jkl}$.

PROOF. We calculate $(c_2a_2^2)(C_2^+ : 1)_{jkl} - (c_1a_1^2)(C_2^- : 1)_{jkl}$, i.e. eliminate the term $c_{j,k+1,l}(a)$. Then,

$$2\sqrt{c_1c_2}a_1a_2(d-j+1)(k+1)\, c_{j-1,k+1,l+1}(a)$$

$$= (r-k)[c_1a_1^2\partial_2 + c_2a_2^2\partial_1 - \frac{1}{2}(r-s)(c_1a_1^2+c_2a_2^2) - \frac{1}{2}u(c_1a_1^2-c_2a_2^2) \\ -(k+l+3)(c_1a_1^2+c_2a_2^2)]\, c_{j,k,l+1}(a). \qquad (147)$$

Substituting l into $l-1$ in (147) and using the equality $2(j+k+l) = d+r+s$, we have the desired $(CJ^-)_{jkl}$.

Next we calculate $(c_1a_1^2)(C_2^+ : 2)_{jkl} - (c_2a_2^2)(C_2^- : 2)_{jkl}$, i.e. eliminate the term $c_{j,k,l+1}(a)$. Then,

$$2\sqrt{c_1c_2}a_1a_2(j+1)(k-r)\, c_{j+1,k,l}(a) \\ = -k[c_1a_1^2\partial_2 + c_2a_2^2\partial_1 + \frac{1}{2}(r-s)(c_1a_1^2+c_2a_2^2) + \frac{1}{2}u(c_1a_1^2-c_2a_2^2) \\ +(k+l-2r-1)(c_1a_1^2+c_2a_2^2)]\, c_{j,k+1,l}(a). \qquad (148)$$

Substituting k into $k-1$ in (148) and using the equality $2(j+k+l) = d+r+s$, then we have the desired $(CJ^+)_{jkl}$.

We compute $(C_2^+ : 1)_{jkl} - (C_2^- : 1)_{jkl}$ and $(C_2^- : 2)_{jkl} - (C_2^+ : 2)_{jkl}$. Then,

$$2(k+1)(l-s)\, c_{j,k+1,l}(a) \\ = (r-k)[\partial_1 + \partial_2 - 4\pi(c_1a_1^2 - c_2a_2^2) - (2l+r-s+4)]\, c_{j,k,l+1}(a), \\ 2(r-k)(l+1)\, c_{j,k,l+1}(a) \\ = -(k+1)[\partial_1 + \partial_2 + 4\pi(c_1a_1^2 - c_2a_2^2) + (2l-r-s-2)]\, c_{j,k+1,l}(a).$$

These equations are $(CK^+L^-)_{jkl}$ and $(CK^-L^+)_{jkl}$.

By using (CK^+L^-) and (CK^-L^+), we eliminate $c_{j,k+1,l}(a)$. Then we have,

$$[\partial_1 + \partial_2 + 4\pi(c_1a_1^2 - c_2a_2^2) + (2l-r-s-2)] \\ \times[\partial_1 + \partial_2 - 4\pi(c_1a_1^2 - c_2a_2^2) - (2l+r-s+4)]\, c_{j,k,l+1}(a) \\ +4(l-s)(l+1)\, c_{j,k,l+1}(a) = 0.$$

Thus we have,

$$\Big[(\partial_1+\partial_2)^2 - 4\pi\{(\partial_1+\partial_2)\triangle\} - 4\pi\triangle(\partial_1+\partial_2) \\ -(2l+r-s+4)(\partial_1+\partial_2) + 4\pi\triangle(\partial_1+\partial_2) - (4\pi\triangle)^2 \\ -(2l+r-s+4)4\pi\triangle + (2l-r-s-2)(\partial_1+\partial_2) \\ -(2l-r-s-2)4\pi\triangle - (2l-r-s-2)(2l+r-s+4) \\ +4(l-s)(l+1)\Big]\, c_{j,k,l+1}(a) \\ = \Big[(\partial_1+\partial_2)^2 - 2(r+3)(\partial_1+\partial_2) + 8\pi(s-2l-2)\triangle - (4\pi\triangle)^2 \\ +4(l-s)(l+1) - (2l-r-s-2)(2l+r-s+4)\Big]\, c_{j,k,l+1}(a).$$

By substituting l into $l-1$, the constant term equals

$$\begin{aligned}&4(l-1-s)l-(2l-r-s-4)(2l+r-s+2)\\&=-4(1+s)l-(-2s-2)2l+(r+s+4)(r-s+2)\\&=(r+3)^2-(s+1)^2.\end{aligned}$$

The rest is clear. □

At first we determine the shape of $c_{0,k,l}(a)$ and $c_{d,k',l'}(a)$ by using only necessary condition.

PROPOSITION 13.9. *We can write $c_{0,k,l}(a)$ as a linear combination of the following two functions.*

$$c^{(1)}_{0,k,l}(a)=(\sqrt{c_1}a_1)^{r+2+\frac{d-u}{2}}(\sqrt{c_2}a_2)^{r+2+\frac{d+u}{2}}(c_1a_1^2-c_2a_2^2)^{-\frac{r+d+2}{2}}\times W_{\frac{s}{2}-l,\frac{s+1}{2}}(4\pi c_1a_1^2-4\pi c_2a_2^2),\tag{149}$$

$$c^{(2)}_{0,k,l}(a)=(\sqrt{c_1}a_1)^{r+2+\frac{d-u}{2}}(\sqrt{c_2}a_2)^{r+2+\frac{d+u}{2}}(c_1a_1^2-c_2a_2^2)^{-\frac{r+d+2}{2}}\times M_{\frac{s}{2}-l,\frac{s+1}{2}}(4\pi c_1a_1^2-4\pi c_2a_2^2).\tag{150}$$

And also $c_{d,k',l'}(a)$ is a linear combination of the following two functions.

$$c^{(1)}_{d,k',l'}(a)=(\sqrt{c_1}a_1)^{r+2+\frac{d+u}{2}}(\sqrt{c_2}a_2)^{r+2+\frac{d-u}{2}}(c_1a_1^2-c_2a_2^2)^{-\frac{r+d+2}{2}}\times W_{\frac{s}{2}-l',\frac{s+1}{2}}(4\pi c_1a_1^2-4\pi c_2a_2^2),\tag{151}$$

$$c^{(2)}_{d,k',l'}(a)=(\sqrt{c_1}a_1)^{r+2+\frac{d+u}{2}}(\sqrt{c_2}a_2)^{r+2+\frac{d-u}{2}}(c_1a_1^2-c_2a_2^2)^{-\frac{r+d+2}{2}}\times M_{\frac{s}{2}-l',\frac{s+1}{2}}(4\pi c_1a_1^2-4\pi c_2a_2^2).\tag{152}$$

Here, $W_{\kappa,\mu}(z)$, $M_{\kappa,\mu}(z)$ are Whittaker's confluent hypergeometric functions. (See [**WW**] *Chapter 16 for definition.)*

PROOF. Put $j=0$ in $(CJ^-)_{jkl}$ in Proposition 13.8, then

$$[2c_2a_2^2\{\partial_1-(r+2+\frac{d-u}{2})\}+2c_1a_1^2\{\partial_2-(r+2+\frac{d+u}{2})\}]\,c_{0,k,l}(a)=0.\tag{153}$$

So if we write $c_{0,k,l}(a_1,a_2)=(\sqrt{c_1}a_1)^{r+2+\frac{d-u}{2}}(\sqrt{c_2}a_2)^{r+2+\frac{d+u}{2}}h_{0,k,l}(a_1,a_2)$ then the functions $h_{0,k,l}(a_1,a_2)$, $(2k+2l=d+r+s)$ satisfy the following differential equations:

$$[2c_2a_2^2\partial_1+2c_1a_1^2\partial_2]\,h_{0,k,l}(a)=0.\tag{154}$$

A solution of (154) can be written as

$$h_{0,k,l}(a_1,a_2)=H_{0,k,l}(4\pi c_1a_1^2-4\pi c_2a_2^2),$$

by using an one variable function $H_{0,k,l}(x)$ on $\mathbb{R}$.

We determine $H_{0,k,l}(x)$. Since $c_{0,k,l}(a)$ satisfies the differential equation $(CS)_{0,k,l}$ in Proposition 13.8, rewrite it into the equation of $h_{0,k,l}(a)$. Then

$$\Big[(\partial_1+r+2+\frac{d-u}{2}+\partial_2+r+2+\frac{d+u}{2})^2$$
$$-2(r+3)(\partial_1+r+2+\frac{d-u}{2}+\partial_2+r+2+\frac{d+u}{2})$$
$$+8\pi(s-2l)\triangle-(4\pi\triangle)^2+(r+3)^2-(s+1)^2\Big]\,h_{0,k,l}(a)=0.$$

Thus we have,

$$[(\partial_1+\partial_2)^2+2(r+d+1)(\partial_1+\partial_2)+8\pi(s-2l)\triangle-(4\pi\triangle)^2$$
$$+(r+d+1)^2-(s+1)^2]\,h_{0,k,l}(a)=0. \tag{155}$$

Next we rewrite (155) into the equation of $H_{0,k,l}(x)$. Put $x=4\pi\triangle=4\pi(c_1a_1^2-c_2a_2^2)$. By using the fact:

$$(\partial_1+\partial_2)H(4\pi\triangle)=\partial_1H(4\pi\triangle)+\partial_2H(4\pi\triangle)$$
$$=(\frac{d}{dx}H)(4\pi\triangle)\cdot4\pi\cdot2c_1a_1^2+(\frac{d}{dx}H)(4\pi\triangle)\cdot4\pi(-2c_2a_2^2)$$
$$=8\pi\triangle(\frac{d}{dx}H)(4\pi\triangle). \tag{156}$$

Therefore we have

$$\Big[4(x\frac{d}{dx})^2+4(r+d+1)x\frac{d}{dx}+2(s-2l)x-x^2$$
$$+(r+d+1)^2-(s+1)^2\Big]H_{0,k,l}(x)=0, \tag{157}$$

i.e.

$$\Big[\frac{d^2}{dx^2}+(r+d+2)\frac{1}{x}\frac{d}{dx}+(\frac{s}{2}-l)\frac{1}{x}-\frac{1}{4}$$
$$+\frac{1}{4x^2}\{(r+d+1)^2-(s+1)^2\}\Big]\,H_{0,k,l}(x)=0. \tag{158}$$

If we write $H_{0,k,l}(x)=x^{-\frac{r+d+2}{2}}F_{0,k,l}(x)$, then

$$\Big[\frac{d^2}{dx^2}-\frac{1}{4}+\frac{\frac{s}{2}-l}{x}+\frac{\frac{1}{4}-(\frac{s+1}{2})^2}{x^2}\Big]\,F_{0,k,l}(x)=0. \tag{159}$$

This is precise Whittaker's confluent hypergeometric differential equation.

Next put $j = d$ in $(CJ^+)_{j,k',l'}$ in Proposition 13.8,

$$[2c_2a_2^2\{\partial_1 - (r+2+\frac{d+u}{2})\} + 2c_1a_1^2\{\partial_2 - (r+2+\frac{d-u}{2})\}]\, c_{d,k',l'}(a) = 0. \tag{160}$$

Then the rest is quite the same with the case of $c_{0,k,l}(a)$. □

PROPOSITION 13.10. *If $d \neq |u|$ then all $c_{j,k,l}(a)$, $(2j+2k+2l = d+r+s)$ are identically zero.*

PROOF. We have the following formula by induction on j (use (153) and $(CJ^+)_{jkl}$ again and again).

$$\begin{aligned}
&2^j j! \binom{k}{j}^{-1} \binom{k-r-1}{j} (-1)^j\, c_{j,k-j,l}(a) \\
&= \Big[(d-u)(d-u-2)\cdots(d-u-2(j-1))(\sqrt{c_1}a_1)^{-j}(\sqrt{c_2}a_2)^j \\
&\quad + j(d-u)\cdots(d-u-2(j-2))\cdot(d+u)(\sqrt{c_1}a_1)^{-j+2}(\sqrt{c_2}a_2)^{j-2} \\
&\quad + \binom{j}{2}(d-u)\cdots(d-u-2(j-3))\cdot(d+u)(d+u-2) \\
&\qquad \times(\sqrt{c_1}a_1)^{-j+4}(\sqrt{c_2}a_2)^{j-4} \\
&\quad \vdots \\
&\quad + j(d-u)\cdot(d+u)(d+u-2)\cdots(d+u-2(j-2)) \\
&\qquad \times(\sqrt{c_1}a_1)^{j-2}(\sqrt{c_2}a_2)^{-j+2} \\
&\quad + (d+u)(d+u-2)\cdots(d+u-2(j-1)) \\
&\qquad \times(\sqrt{c_1}a_1)^{j}(\sqrt{c_2}a_2)^{-j}\Big]\, c_{0,k,l}(a).
\end{aligned}$$

Put $j = d$ in the above formula, Proposition 13.9 says that d must be either u or $-u$, which proves the proposition. □

13.4. Main theorem for definite characters.

THEOREM 13.11. *For $\pi = \pi_\Lambda^*$, $\Lambda = [r-1, s+1; u] \in \Xi_{\rm III}$, $(r-s-2 > |u|)$, a positive definite character ξ of N_S and an irreducible finite-dimensional representation χ of $SU(\xi)$ $(\simeq SU(2))$ with* $\dim\chi = d+1$, *we have*

$$\dim_{\mathbb{C}} \mathrm{Hom}_{(\mathfrak{g},K)}(\pi_\Lambda^*, C_\eta^\infty(R\backslash G)) = 0. \tag{161}$$

PROOF. We may assume that $d = |u|$. If not so, the system of partial differential equations for coefficient functions $c_{jkl}(a)$ in $\Phi_{\pi,\tau}(a)$ have no non-zero solution. (Propositions 13.7 and 13.10). When $d = |u|$, $c_{0,k,l}(a)$ has a pole at $\{c_1a_1^2 - c_2a_2^2 = 0\}$ because c_1 and c_2 are positive and is not analytic (Proposition 13.9). So we completed the proof. □

14. Generalized Whittaker functions for the middle discrete series associated with indefinite H_ξ

14.1. Explicit differential equations. We write explicit differential equations of generalized Whittaker functions $\Phi_{\pi,\tau}(a)$ for the case: $\lceil \quad \pi = \pi^*_\Lambda,\ \Lambda \in \Xi_{\mathrm{III}}$ with their minimal K-type τ^*, $\tau = [r,s;u] \quad \rfloor$.

Let $\Phi_{\pi,\tau}(a) = \sum_{j\in\mathbb{Z}_\epsilon}\sum_{\substack{0\le k\le r\\ 0\le l\le s}} b_{jkl}(a)(u_j\otimes f_{kl}) \in C^\infty_{M_0}(A; V_\chi\otimes W_\tau)$. Here, $\eta = \chi\otimes\xi$, $\chi = \chi_{\epsilon,\mu}$ is (an irreducible) unitary representation of $SU(\xi)\simeq SU(1,1)$ and ξ is a nondegenerate unitary character of N_S determined by an indefinite matrix $H_\xi = \begin{pmatrix} c_1 & \\ & c_2 \end{pmatrix}$. We assume that $c_1 > 0$ and $c_2 < 0$.

Then the unknown functions $b_{jkl}(a)$ satisfy the following difference differential equations.

PROPOSITION 14.1. *The coefficient functions $b_{jkl}(a)$ in $\Phi_{\pi,\tau}(a)$ satisfy the following system of partial differential equations. Here, we put $\partial_i = a_i\frac{\partial}{\partial a_i}$: the Euler operator $(i=1,2)$ and $\triangle = (c_1a_1^2 - c_2a_2^2)$.*

$$
\begin{aligned}
&(P^{(-,-)})_{jkl}\\
&\quad \sqrt{-c_1c_2}\frac{a_1a_2}{\triangle}\frac{1}{2}(j-1+\mu)(k+1)(l+1)\, b_{j-2,k+1,l+1}(a)\\
&\quad +\sqrt{-c_1c_2}\frac{a_1a_2}{\triangle}\frac{1}{2}(-j-1+\mu)(k-r)(s-l)\, b_{j+2,k,l}(a)\\
&\quad +\frac{1}{2}(k-r)(l+1)\Big[\partial_1 - 4\pi c_1a_1^2 + (b_1-k+l-2s-3)\\
&\qquad\qquad +(s-l+k+1)\frac{2c_1a_1^2}{\triangle}\Big]\, b_{j,k,l+1}(a)\\
&\quad +\frac{1}{2}(k+1)(s-l)\Big[\partial_2 - 4\pi c_2a_2^2 + (b_2+k-2r-l-3)\\
&\qquad\qquad -(r-k+l+1)\frac{2c_2a_2^2}{\triangle}\Big]\, b_{j,k+1,l}(a)\\
&\quad = 0 \qquad (0\le k\le r-1,\ 0\le l\le s-1,\ j = r+s-2k-2l-2),
\end{aligned}
$$

$$
\begin{aligned}
&(P^{(-,+)})_{jkl}\\
&\quad -\sqrt{-c_1c_2}\frac{a_1a_2}{\triangle}\frac{1}{2}(j-1+\mu)(k+1)\, b_{j-2,k+1,l}(a)\\
&\quad +\sqrt{-c_1c_2}\frac{a_1a_2}{\triangle}\frac{1}{2}(-j-1+\mu)(k-r)\, b_{j+2,k,l-1}(a)
\end{aligned}
$$

$$
\begin{aligned}
&+\frac{1}{2}(r-k)\Big[\partial_1 - 4\pi c_1 a_1^2 + (b_1 - k + l - 2) \\
&\qquad\qquad +(1+k-l)\frac{2c_1a_1^2}{\triangle}\Big]\, b_{j,k,l}(a) \\
&+\frac{1}{2}(k+1)\Big[\partial_2 - 4\pi c_2 a_2^2 + (b_2 + k - l - 2r + 2s) \\
&\qquad\qquad +(1+k-l-r+s)\frac{2c_2a_2^2}{\triangle}\Big]\, b_{j,k+1,l-1}(a) \\
&= 0 \qquad (0 \le k \le r-1,\ 0 \le l \le s+1,\ j = r+s-2k-2l),
\end{aligned}
$$

$$
\begin{aligned}
&(\bar{P}^{(-,-)})_{jkl} \\
&\quad \sqrt{-c_1c_2}\frac{a_1a_2}{\triangle}\frac{1}{2}(j-1+\mu)(k+1)(l+1)\, b_{j-2,k+1,l+1}(a) \\
&\quad +\sqrt{-c_1c_2}\frac{a_1a_2}{\triangle}\frac{1}{2}(-j-1+\mu)(k-r)(s-l)\, b_{j+2,k,l}(a) \\
&\quad +\frac{1}{2}(k+1)(l-s)\Big[\partial_1 + 4\pi c_1 a_1^2 - (b_1 - k + l + 2r + 3) \\
&\qquad\qquad +(r-k+l+1)\frac{2c_1a_1^2}{\triangle}\Big]\, b_{j,k+1,l}(a) \\
&\quad +\frac{1}{2}(r-k)(l+1)\Big[\partial_2 + 4\pi c_2 a_2^2 - (b_2 + k - l + 2s + 3) \\
&\qquad\qquad -(k+s-l+1)\frac{2c_2a_2^2}{\triangle}\Big]\, b_{j,k,l+1}(a) \\
&\quad = 0 \qquad (0 \le k \le r-1,\ 0 \le l \le s-1,\ j = r+s-2k-2l-2),
\end{aligned}
$$

$$
\begin{aligned}
&(\bar{P}^{(-,+)})_{jkl} \\
&\quad -\sqrt{-c_1c_2}\frac{a_1a_2}{\triangle}\frac{1}{2}(j-1+\mu)(k+1)\, b_{j-2,k+1,l}(a) \\
&\quad +\sqrt{-c_1c_2}\frac{a_1a_2}{\triangle}\frac{1}{2}(-j-1+\mu)(k-r)\, b_{j+2,k,l-1}(a) \\
&\quad -\frac{1}{2}(k+1)\Big[\partial_1 + 4\pi c_1 a_1^2 - (b_1 - k + l + 2r - 2s) \\
&\qquad\qquad +(r-k-s+l-1)\frac{2c_1a_1^2}{\triangle}\Big]\, b_{j,k+1,l-1}(a) \\
&\quad +\frac{1}{2}(k-r)\Big[\partial_2 + 4\pi c_2 a_2^2 - (b_2 + k - l + 2)
\end{aligned}
$$

$$-(k-l+1)\frac{2c_2a_2^2}{\triangle}\Big]\, b_{j,k,l}(a)$$

$$= 0 \qquad (0 \le k \le r-1,\ 0 \le l \le s+1,\ j = r+s-2k-2l).$$

PROOF. By using the action of $SU(\xi) \simeq SU(1,1)$ in Proposition 5.2, the proof is quite similar with the definite case. (See Proposition 13.1.) □

We firstly reduce the system to the following.

PROPOSITION 14.2. *The functions $b_{jkl}(a)$ satisfy the following system of partial differential equations.*

$(B_2^+ : 1)_{jkl}$

$$\sqrt{-c_1c_2}\frac{a_1a_2}{\triangle}\frac{1}{2}(j-1+\mu)(k+1)\, b_{j-2,k+1,l+1}(a)$$

$$+(k+1)(l-s)\frac{c_1a_1^2}{\triangle}\, b_{j,k+1,l}(a)$$

$$+\frac{1}{2}(k-r)\Big[\partial_1 - 4\pi c_1a_1^2 + (b_1-k-l-3)$$

$$+(k+1)\frac{2c_1a_1^2}{\triangle}\Big]\, b_{j,k,l+1}(a)$$

$$= 0 \qquad (0 \le k \le r-1,\ -1 \le l \le s-1,$$
$$j = r+s-2k-2l-2),$$

$(B_2^+ : 2)_{jkl}$

$$\sqrt{-c_1c_2}\frac{a_1a_2}{\triangle}\frac{1}{2}(-j-1+\mu)(k-r)\, b_{j+2,k,l}(a)$$

$$+(k-r)(l+1)\frac{c_2a_2^2}{\triangle}\, b_{j,k,l+1}(a)$$

$$+\frac{1}{2}(k+1)\Big[\partial_2 - 4\pi c_2a_2^2 + (b_2+k+l-2r-1)$$

$$-(r-k)\frac{2c_2a_2^2}{\triangle}\Big]\, b_{j,k+1,l}(a)$$

$$= 0 \qquad (0 \le k \le r-1,\ 0 \le l \le s,\ j = r+s-2k-2l-2),$$

$(B_2^- : 1)_{jkl}$

$$\sqrt{-c_1c_2}\frac{a_1a_2}{\triangle}\frac{1}{2}(j-1+\mu)(k+1)\, b_{j-2,k+1,l+1}(a)$$

$$
\begin{aligned}
&+(k+1)(l-s)\frac{c_2a_2^2}{\triangle}\, b_{j,k+1,l}(a)\\
&+\frac{1}{2}(r-k)\Big[\partial_2+4\pi c_2a_2^2-(b_2+k+l+3)\\
&\qquad\qquad -(k+1)\frac{2c_2a_2^2}{\triangle}\Big]\, b_{j,k,l+1}(a)\\
&=0 \qquad (0\le k\le r-1,\ -1\le l\le s-1,\\
&\qquad\qquad j=r+s-2k-2l-2),
\end{aligned}
$$

$(B_2^- : 2)_{jkl}$

$$
\begin{aligned}
&\sqrt{-c_1c_2}\frac{a_1a_2}{\triangle}\frac{1}{2}(-j-1+\mu)(k-r)\, b_{j+2,k,l}(a)\\
&+(k-r)(l+1)\frac{c_1a_1^2}{\triangle}\, b_{j,k,l+1}(a)\\
&-\frac{1}{2}(k+1)\Big[\partial_1+4\pi c_1a_1^2-(b_1-k-l+2r+1)\\
&\qquad\qquad +(r-k)\frac{2c_1a_1^2}{\triangle}\Big]\, b_{j,k+1,l}(a)\\
&=0 \qquad (0\le k\le r-1,\ 0\le l\le s,\ j=r+s-2k-2l-2).
\end{aligned}
$$

PROOF. The equation $(B_2^- : 1)_{jkl}$ is obtained from:

$$(\bar{P}^{(-,-)})_{j,k,l}-(s-l)(\bar{P}^{(-,+)})_{j,k,l+1}.$$

The equation $(B_2^- : 2)_{jkl}$ is obtained from:

$$(\bar{P}^{(-,-)})_{j,k,l}+(l+1)(\bar{P}^{(-,+)})_{j,k,l+1}.$$

The equations $(B_2^+ : 1)$ and $(B_2^+ : 2)$ are obtained from $(P^{-,-})_{j,k,l}$ and $(P^{-,+})_{j,k,l+1}$ similarly. □

14.2. The case of holonomic system of rank 0.

PROPOSITION 14.3. *If χ is a unitary principal series or complementary series representation then the coefficient functions $b_{jkl}(a)$ ($j\in\mathbb{Z}_\epsilon$, $0\le k\le r$, $0\le l\le s$ and $j+2k+2l=r+s$), which satisfy the partial differential equations $(B_2^+ : 1)_{jkl},\dots,(B_2^- : 2)_{jkl}$ in Proposition 14.2, are identically zero.*

PROOF. The assumption on χ implies that j satisfying $j+2k+2l=r+s$ always exist for all k,l. So we write $b_{kl}(a)=b_{r+s-2k-2l,kl}(a)$ for simplicity in this subsection. We will show all $b_{kl}(a)\equiv 0$ in this case.

From $(B_2^+ : 1) - (B_2^- : 1)$ and $(B_2^- : 2) - (B_2^+ : 1)$, we have

$(C)_{kl}$

$$\begin{aligned} &2(k+1)(l-s)\, b_{k+1,l}(a) \\ &= (r-k)[\partial_1 + \partial_2 - 4\pi(c_1 a_1^2 - c_2 a_2^2) - (r-s+2l+4)]\, b_{k,l+1}(a), \end{aligned}$$

and

$(D)_{kl}$

$$\begin{aligned} &2(k-r)(l+1)\, b_{k,l+1}(a) \\ &= (k+1)[\partial_1 + \partial_2 + 4\pi(c_1 a_1^2 - c_2 a_2^2) - (r+s-2l+2)]\, b_{k+1,l}(a). \end{aligned}$$

These equations are identical with $(A)_{kl}$ and $(B)_{kl}$ in the proof of Proposition 13.3. So we have $b_{kl}(a) \equiv 0$ $(0 \le k \le r, 0 \le l \le s)$. $\square$

14.3. The case of holonomic system of rank 1. We assume that χ is one of the following representations (see Proposition 5.3).

(1) Trivial representation 1.

(2),(3) (Limits of) the discrete series representations,

$\chi = D_p^+$ with representation space $V_p = \bigoplus_{j \in p+2\mathbb{Z}_{\ge 0}} \mathbb{C} u_j$, and

$\chi = D_p^-$ with representation space $\bar{V}_p = \bigoplus_{j \in -p+2\mathbb{Z}_{\le 0}} \mathbb{C} u_j$, for $p \ge 1$.

PROPOSITION 14.4. *Let χ be one of D_p^+, D_p^- and 1 and μ be $p-1$, $p-1$, 0 respectively. The functions $b_{jkl}(a)$ satisfy the following system of partial differential equations. This system is equivalent to the that of Proposition 14.2.*

$(BJ^-)_{jkl}$

$$\begin{aligned} &-2\sqrt{-c_1 c_2}\, a_1 a_2 (j-1+\mu)(k+1)\, b_{j-2,k+1,l}(a) \\ &= (r-k)\Big[2c_2 a_2^2 \partial_1 - 2c_2 a_2^2 (r+2-\frac{u+j}{2}) \\ &\qquad + 2c_1 a_1^2 \partial_2 - 2c_1 a_1^2 (r+2+\frac{u-j}{2})\Big]\, b_{j,k,l}(a) \\ &\qquad (0 \le k \le r-1,\ 0 \le l \le s,\ j = r+s-2k-2l), \end{aligned}$$

$(BJ^+)_{jkl}$

$$\begin{aligned} &2\sqrt{-c_1 c_2}\, a_1 a_2 (-j-1+\mu)(k-r-1)\, b_{j+2,k-1,l}(a) \\ &= -k\Big[2c_2 a_2^2 \partial_1 - 2c_2 a_2^2 (r+2+\frac{u+j}{2}) \\ &\qquad + 2c_1 a_1^2 \partial_2 - 2c_1 a_1^2 (r+2-\frac{u-j}{2})\Big]\, b_{j,k,l}(a) \\ &\qquad (1 \le k \le r,\ 0 \le l \le s,\ j = r+s-2k-2l), \end{aligned}$$

$(BK^+L^-)_{jkl}$

$$\begin{aligned}
&2(k+1)(l-s)\, b_{j,k+1,l}(a) \\
&= (r-k)\Big[\partial_1+\partial_2 - 4\pi(c_1a_1^2-c_2a_2^2) \\
&\qquad -(r-s+2l+4)\Big]\, b_{j,k,l+1}(a) \\
&\quad (0\le k\le r-1,\ -1\le l\le s-1, \\
&\qquad j=r+s-2k-2l-2),
\end{aligned}$$

$(BK^-L^+)_{jkl}$

$$\begin{aligned}
&2(k-r)(l+1)\, b_{j,k,l+1}(a) \\
&= (k+1)\Big[\partial_1+\partial_2 + 4\pi(c_1a_1^2-c_2a_2^2) \\
&\qquad -(r+s-2l+2)\Big]\, b_{j,k+1,l}(a) \\
&\quad (0\le k\le r-1,\ 0\le l\le s, \\
&\qquad j=r+s-2k-2l-2),
\end{aligned}$$

$(BS)_{jkl}$

$$\begin{aligned}
&\Big[(\partial_1+\partial_2)^2 - 2(r+3)(\partial_1+\partial_2) + 8\pi(s-2l)(c_1a_1^2-c_2a_2^2) \\
&-16\pi^2(c_1a_1^2-c_2a_2^2)^2 + (r+3)^2 - (s+1)^2\Big]\, b_{j,k,l}(a) = 0.
\end{aligned}$$

Remark. $(BK^+L^-)_{jkl}$ and $(BK^-L^+)_{jkl}$ imply $(BS)_{jkl}$.

PROOF. Similar to the definite case (See Proposition 13.8). □

The equations $(BK^+L^-)_{jkl}$ and $(BK^-L^+)_{jkl}$ in Proposition 14.4 implies the following formulas.

LEMMA 14.5. *Fix j, k_1 and k_2 such that $j+2k_1=r+s$ and so $j+2k_2+2s=r+s$ respectively. Then we have the following formula. For $0\le w_1\le \min(k_1,s)$,*

$$(-1)^{w_1}\binom{r-k_1+w_1}{w_1}(w_1)!\, b_{j,k_1-w_1,w_1}(a) = \binom{k_1}{w_1}\mathcal{F}^-_{w_1}(4\pi\triangle)\, b_{j,k_1,0}(a). \tag{162}$$

For $0\le w_2\le \min(r-k_2,s)$,

$$\binom{k_2+w_2}{w_2}(w_2)!\, b_{j,k_2+w_2,s-w_2}(a) = \binom{r-k_2}{w_2}\mathcal{F}^+_{w_2}(4\pi\triangle)\, b_{j,k_2,s}(a). \tag{163}$$

Here, polynomials $\mathcal{F}^\pm_w(x)$ are defined in Definition 13.4 and $\triangle = c_1a_1^2 - c_2a_2^2$.

PROOF. This is essentially the same as Lemma 13.6 in the section for definite characters. □

Firstly, we treat the case of $\chi = D_p^\pm$ with certain condition on p.

PROPOSITION 14.6. *If $p > r+s$ or $p < r-s$, then all $b_{jkl}(a)$ which satisfy the partial differential equations in Proposition 14.4, are identically zero.*

PROOF. If some $b_{jkl}(a)$ is not zero then indices j,k,l must satisfy

$$j + 2k + 2l = r + s. \tag{164}$$

We prove the case of $\chi = D_p^+$. The D_p^- case is proved quite similar manner.

(1) the case $p > r+s$: There is no (j,k,l) which satisfy (164), because the condition $p > r+s$ implies $j > r+s$.

(2) the case $p < r-s$: We can write $p = r - s - 2\delta$ by using an integer δ such that $0 < 2\delta < r-s$. Put $j = p$ in (164), then we have $k + l = s + \delta$. Thus

$$(k,l) = (s+\delta, 0), (s+\delta-1, 1), \ldots, (\delta, s). \tag{165}$$

Note that $s + \delta < r$. We firstly put $j = p$ and $(k,l) = (s+\delta, 0)$ in (BJ^-), and also put $j = p$, $(k, l+1) = (s+\delta, 0)$ in (BK^+L^-). Then, $b_{p,s+\delta,0}(a)$ satisfies the following two partial differential equations,

$$\Big[2c_2a_2^2\partial_1 - 2c_2a_2^2(r + 2 - \frac{p+u}{2})$$
$$+2c_1a_1^2\partial_2 - 2c_1a_1^2(r+2-\frac{p-u}{2})\Big]\, b_{p,s+\delta,0}(a) = 0,$$
$$\Big[\partial_1 + \partial_2 - 4\pi(c_1a_1^2 - c_2a_2^2) - (r-s+2)\Big]\, b_{p,s+\delta,0}(a) = 0.$$

Let us solve them. If we write $b_{p,s+\delta,0}(a) = a_1^{r+2-\frac{p+u}{2}} a_2^{r+2-\frac{p-u}{2}} g(a)$, then $g(a)$ satisfies,

$$[2c_2a_2^2\partial_1 + 2c_1a_1^2\partial_2]\, g(a) = 0, \tag{166}$$
$$[\partial_1 + \partial_2 - 4\pi(c_1a_1^2 - c_2a_2^2) + (r+s-p+2)]\, g(a) = 0. \tag{167}$$

Any solution of (166) can be written as $g(a) = G(4\pi\triangle)$ by using a one variable function $G(x)$ on $\mathbb{R}$. Here, $\triangle = c_1a_1^2 - c_2a_2^2$. Note that $(\partial_1 + \partial_2)\triangle = 2\triangle$, and rewrite (167) of $G(x)$. Then we have,

$$\left[2x\frac{d}{dx} - x + (r+s-p+2)\right] G(x) = 0.$$

Thus $G(x) = x^{-\frac{r+s-p+2}{2}} e^{\frac{x}{2}}$ up to a constant multiple.

Therefore, $b_{p,s+\delta,0}(a)$ is given by up to constant

$$b_{p,s+\delta,0}(a) = (\sqrt{c_1}a_1)^{r+2-\frac{p+u}{2}}(\sqrt{-c_2}a_2)^{r+2-\frac{p-u}{2}} \times (c_1a_1^2 - c_2a_2^2)^{-\frac{r+s-p+2}{2}} e^{2\pi(c_1a_1^2-c_2a_2^2)}. \tag{168}$$

We remark that this function is analytic on $A \simeq (\mathbb{R}_{>0})^2$ and is *exponentially increasing* when each a_1, a_2 tends to infinity from the assumption $c_1 > 0$ and $c_2 < 0$.

Next put $j = p$ and $(k, l) = (\delta, s)$ in (BJ^-), and also put $j = p$, $(k + 1, l) = (\delta, s)$ in (BK^-L^+). Then, $b_{p,\delta,s}(a)$ satisfies the following two equations,

$$\Big[2c_2a_2^2\partial_1 - 2c_2a_2^2(r + 2 - \frac{p+u}{2}) + 2c_1a_1^2\partial_2 - 2c_1a_1^2(r + 2 - \frac{p-u}{2})\Big]\, b_{p,\delta,s}(a) = 0, \tag{169}$$

$$\Big[\partial_1 + \partial_2 + 4\pi(c_1a_1^2 - c_2a_2^2) - (r - s + 2)\Big]\, b_{p,\delta,s}(a) = 0. \tag{170}$$

We solve them like the previous two equations. Then $b_{p,\delta,s}(a)$ is determined by up to constant

$$b_{p,\delta,s}(a) = (\sqrt{c_1}a_1)^{r+2-\frac{p+u}{2}}(\sqrt{-c_2}a_2)^{r+2-\frac{p-u}{2}} \times (c_1a_1^2 - c_2a_2^2)^{-\frac{r+s-p+2}{2}} e^{-2\pi(c_1a_1^2-c_2a_2^2)}. \tag{171}$$

We remark that this function is analytic on $A \simeq (\mathbb{R}_{>0})^2$ and *rapidly decreasing* when each a_1, a_2 tends to infinity from the assumption $c_1 > 0$ and $c_2 < 0$.

Let us use Lemma 14.5. Put $j = p$, $k_1 = s + \delta$ and $k_2 = \delta$ in the lemma. Then we have,
for $0 \leq w_1 \leq s$,

$$(-1)^{w_1}\binom{r-s-\delta+w_1}{w_1}(w_1)!\, b_{p,s+\delta-w_1,w_1}(a) = \binom{s+\delta}{w_1}\mathcal{F}^-_{w_1}(4\pi\triangle)\, b_{p,s+\delta,0}(a), \tag{172}$$

and for $0 \leq w_2 \leq s$,

$$\binom{\delta+w_2}{w_2}(w_2)!\, b_{p,\delta+w_2,s-w_2}(a) = \binom{r-\delta}{w_2}\mathcal{F}^+_{w_2}(4\pi\triangle)\, b_{p,\delta,s}(a). \tag{173}$$

Therefore, the functions

$$b_{p,s+\delta,0}(a), b_{p,s+\delta-1,1}(a), \ldots, b_{p,\delta,s}(a),$$

are equal to each other up to *polynomial* functions. We use $b_{p,s+\delta,0}(a)$ and $b_{p,\delta,s}(a)$ are exponentially increasing and decreasing respectively, comparing (168) with (171). Thus we have

$$b_{p,s+\delta,0}(a) = b_{p,s+\delta-1,1}(a) = \cdots = b_{p,\delta,s}(a) \equiv 0. \tag{174}$$

We can see easily (BJ^+) and (174) implies that all $b_{jkl}(a)$ are zero. □

When χ is trivial, we have the following.

PROPOSITION 14.7. *If χ is the trivial representation, then all $b_{0kl}(a)$ are identically zero.*

PROOF. If some $b_{0kl}(a)$ is not zero then indices k,l must satisfy

$$2k+2l = r+s.$$

In other words,

$$(k,l) = \left(\frac{r+s}{2},0\right), \left(\frac{r+s}{2}-1,1\right), \ldots, \left(\frac{r-s}{2},s\right),$$

and $(r+s)/2 < r$, $(r-s)/2 > 0$. Since $r-s-2 > |u|$, the rest is the same as the proof of Proposition 14.6. □

In the following, we study the case of the discrete series representations: $\chi = D_p^+$ with $V_p = \bigoplus_{j\in p+2\mathbb{Z}_{\geq 0}} \mathbb{C}u_j$, and $\chi = D_p^-$ with $\bar{V}_p = \bigoplus_{j\in -p+2\mathbb{Z}_{\leq 0}} \mathbb{C}u_j$ with $r-s \leq p \leq r+s$. We require a lemma on the indices (j,k,l) of possibly non-zero $b_{jkl}(a)$.

LEMMA 14.8. *Set $\chi = D_p^\pm$ with $r-s\leq p\leq r+s$. If we write $p = r-s+2q$ by using an integer q such that $0\leq q\leq s$, then the triples listed below are all (j,k,l) which satisfy $j+2k+2l = r+s$.*
When $\chi = D_p^+$

$$\begin{aligned}
&j = r+s, && (k,l) = (0,0),\\
&j = r+s-2, && (k,l) = (1,0),(0,1),\\
&\vdots\\
&j = r-s+2q = p, && (k,l) = (s-q,0),\ldots,(0,s-q).
\end{aligned} \tag{175}$$

When $\chi = D_p^-$,

$$\begin{aligned}
&j = -r-s, && (k,l) = (r,s),\\
&j = -r-s+2, && (k,l) = (r-1,s),(r,s-1),\\
&\vdots\\
&j = -r+s-2q = -p, && (k,l) = (r-s+q,s),\ldots,(r,q).
\end{aligned} \tag{176}$$

PROOF. *By direct calculation.* □

The above lemma leads the following.

LEMMA 14.9. *Put* $l = -1$ *in* (BK^+L^-) *and* $l = s$ *in* (BK^-L^+), *then we have,*

$$(r-k)[\partial_1 + \partial_2 - 4\pi(c_1 a_1^2 - c_2 a_2^2) - (r-s+2)]\, b_{j,k,0}(a) = 0, \quad (177)$$
$$k[\partial_1 + \partial_2 + 4\pi(c_1 a_1^2 - c_2 a_2^2) - (r-s+2)]\, b_{j,k,s}(a) = 0. \quad (178)$$

There is no (j,k,l) *in the list (175) of* D_p^+ *which appear the equation (178) and* $k \neq 0$. *And there is no* (j,k,l) *in the list (176) of* D_p^- *which appear the equation (177) and* $k \neq r$.

PROOF. We may assume $q = 0$. The only possibility that $b_{j,k,l}(a)$ satisfy (178) is the case $k = 0$, which make (178) vanish. The rest is similar. □

Therefore, all $b_{jkl}(a)$ are rapidly increasing or decreasing at infinity according to that $\chi = D_p^+$ or D_p^- respectively. First of all let us determine $b_{p,k,l}(a)$ (resp. $b_{-p,k,l}(a)$), then by using (BJ^+) (resp. (BJ^-)) determine all $b_{jkl}(a)$.

PROPOSITION 14.10. *We put* $p = r - s + 2q$, *for* $q = 0, \dots, s$. *For* $\chi = D_p^+$, *the functions* $b_{p,s-q-w,w}(a)$ $(w = 0, \dots, s-q)$ *in the bottom row of the list (175) in Lemma 14.8 are up to a constant as follows.*

$$\begin{aligned} b_{p,s-q-w,w}(a) &= (-1)^w \binom{r-s+q+w}{w}^{-1} (w!)^{-1} \binom{s-q}{w} \\ &\quad \times \mathcal{F}_w^-(4\pi c_1 a_1^2 - 4\pi c_2 a_2^2)\, b_{p,s-q,0}(a) \\ &= (-1)^w \binom{r-s+q+w}{w}^{-1} (w!)^{-1} \binom{s-q}{w} \\ &\quad \times \mathcal{F}_w^-(4\pi c_1 a_1^2 - 4\pi c_2 a_2^2) \\ &\quad \times (\sqrt{c_1}a_1)^{r+2-b_2-q} (\sqrt{-c_2}a_2)^{r+2+b_1-q} \\ &\quad \times (c_1 a_1^2 - c_2 a_2^2)^{-(1+s-q)} e^{2\pi(c_1 a_1^2 - c_2 a_2^2)}. \end{aligned} \quad (179)$$

These functions are all analytic and rapidly increasing since $c_1 > 0$ *and* $c_2 < 0$.

For $\chi = D_p^-$, *the functions* $b_{-p,r-s+q+w,s-w}(a)$ $(w = 0, \dots, s-q)$ *in the bottom row of the list (176) in Lemma 14.8 are up to a constant as follows.*

$$\begin{aligned} b_{-p,r-s+q+w,s-w}(a) &= \binom{r-s+q+w}{w}^{-1} (w!)^{-1} \binom{s-q}{w} \\ &\quad \times \mathcal{F}_w^+(4\pi c_1 a_1^2 - 4\pi c_2 a_2^2)\, b_{-p,r-s+q,s}(a) \\ &= \binom{r-s+q+w}{w}^{-1} (w!)^{-1} \binom{s-q}{w} \end{aligned}$$

$$\begin{aligned}&\times\mathcal{F}_w^{+}(4\pi c_1a_1^2-4\pi c_2a_2^2)\\&\times(\sqrt{c_1}a_1)^{r+2+b_1-q}(\sqrt{-c_2}a_2)^{r+2-b_2-q}\\&\times(c_1a_1^2-c_2a_2^2)^{-(1+s-q)}e^{-2\pi(c_1a_1^2-c_2a_2^2)}.\end{aligned}\tag{180}$$

These functions are all analytic and rapidly decreasing since $c_1>0$ and $c_2<0$.
Here, polynomials $\mathcal{F}_w^{\pm}(x)$ are defined in Definition 13.4.

PROOF. It suffices to show (180) for $b_{-p,r-s+q,s}$ thanks to Lemma 14.5. We can obtain (179) by similar method. Put $j=-p$, $k=r-s+q$ in (BJ^+) and (178) in Lemma 14.9 and use $r-s+q\neq 0$,

$$\Big[2c_2a_2^2\partial_1-2c_2a_2^2(r+2-\frac{p-u}{2})$$
$$+2c_1a_1^2\partial_2-2c_1a_1^2(r+2-\frac{p+u}{2})\Big]\,b_{-p,r-s+q,s}(a)=0.$$
$$\Big[\partial_1+\partial_2+4\pi(c_1a_1^2-c_2a_2^2)-(r-s+2)\Big]\,b_{-p,r-s+q,s}(a)=0.$$

We have already solved these two partial differential equations in the proof of Proposition 14.6. □

We introduce the following Laurent polynomial of two variables x,y (or the function of x/y). This function is useful to describe $b_{jkl}(a)$ when $j>p$ (resp. $j<-p$).

DEFINITION 14.11 ($\mathcal{R}_h^{(\alpha,\beta)}(x,y)$). For $h=0,\dots,(s-q)$ and two integers α, β such that $\alpha\pm\beta\geq 2$ and $\alpha\equiv\beta \pmod 2$, define the Laurent polynomial by

$$\mathcal{R}_h^{(\alpha,\beta)}(x,y)=\sum_{i=0}^{h}(-1)^i\binom{\frac{\alpha-\beta}{2}+h-i-1}{h-i}\binom{\frac{\alpha+\beta}{2}+i-1}{i}x^{h-2i}y^{-h+2i}.$$

We note the following fact on the index h of $\mathcal{R}_h^{(\alpha,\beta)}(x,y)$.

LEMMA 14.12.

$$\begin{aligned}&\left[\frac{x}{y}\left\{y\frac{\partial}{\partial y}-(h+\alpha-\beta)\right\}-\frac{y}{x}\left\{x\frac{\partial}{\partial x}-(h+\alpha+\beta)\right\}\right]\,\mathcal{R}_h^{(\alpha,\beta)}(x,y)\\&\quad=-2(h+1)\,\mathcal{R}_{h+1}^{(\alpha,\beta)}(x,y).\end{aligned}\tag{181}$$

PROOF. We define two partial differential operators P_1, P_2 appearing in the lemma by

$$P_1=\ \ xy^{-1}\left\{y\frac{\partial}{\partial y}-(h+\alpha-\beta)\right\},$$
$$P_2=-x^{-1}y\left\{x\frac{\partial}{\partial x}-(h+\alpha+\beta)\right\}.$$

Then

$$
\begin{aligned}
&P_1\,\mathcal{R}_h^{(\alpha,\beta)}(x,y)\\
&=\sum_{i=0}^{h}(-1)^i\{-h+2i-h-\alpha+\beta\}\binom{\frac{\alpha-\beta}{2}+h-i-1}{h-i}\\
&\quad\times\binom{\frac{\alpha+\beta}{2}+i-1}{i}(x/y)^{h+1-2i}\\
&=-2\sum_{i=0}^{h}(-1)^i(h-i+1)\binom{\frac{\alpha-\beta}{2}+h-i}{h+1-i}\binom{\frac{\alpha+\beta}{2}+i-1}{i}(x/y)^{h+1-2i}.
\end{aligned}
$$

We have similarly,

$$
\begin{aligned}
&P_2\,\mathcal{R}_h^{(\alpha,\beta)}(x,y)\\
&=-\sum_{i=0}^{h}(-1)^i\{h-2i-h-\alpha-\beta\}\binom{\frac{\alpha-\beta}{2}+h-i-1}{h-i}\\
&\quad\times\binom{\frac{\alpha+\beta}{2}+i-1}{i}(x/y)^{h-1-2i}\\
&=2\sum_{i=0}^{h}(-1)^i(i+1)\binom{\frac{\alpha-\beta}{2}+h-i-1}{h-i}\binom{\frac{\alpha+\beta}{2}+i}{i+1}(x/y)^{h-1-2i}.
\end{aligned}
$$

Thus we get the desired formula.

$$
\begin{aligned}
&(P_1+P_2)\,\mathcal{R}_h^{(\alpha,\beta)}(x,y)\\
&=-2\sum_{i=0}^{h+1}(-1)^i\{(h-i+1)+i\}\binom{\frac{\alpha-\beta}{2}+(h+1)-i-1}{(h+1)-i}\\
&\quad\times\binom{\frac{\alpha+\beta}{2}+i-1}{i}(x/y)^{(h+1)-2i}.\\
&=-2(h+1)\mathcal{R}_{h+1}^{(\alpha,\beta)}(x,y). \quad\square
\end{aligned}
$$

PROPOSITION 14.13. *We put $p=r-s+2q$, for $q=0,\dots,s$. For $\chi=D_p^+$, the functions $b_{p+2h,k-h,l}(a)$ $(h=0,\dots,k)$ in the $(l+1)$-th column of the list (175) in Lemma 14.8 are as follows.*

$$
b_{p+2h,k-h,l}(a)=\binom{r-k+h}{h}^{-1}\binom{k}{h}\mathcal{R}_h^{(p,u)}(\sqrt{c_1}a_1,\sqrt{-c_2}a_2)\,b_{p,k,l}(a). \tag{182}
$$

For $\chi = D_p^-$, the functions $b_{-p-2h,k+h,l}(a)$ $(h = 0, \dots, r-k)$ in the $(l+1)$-th column of the list (176) in Lemma 14.8 are as follows.

$$b_{-p-2h,k+h,l}(a) = \binom{k+h}{h}^{-1} \binom{r-k}{h} \mathcal{R}_h^{(p,u)}(\sqrt{-c_2}a_2, \sqrt{c_1}a_1)\, b_{-p,k,l}(a). \tag{183}$$

PROOF. We show (183) by induction on h. The equation (182) is obtained on a parallel with it. The case $h = 0$ is trivial. Let $h > 0$, we substitute (183) into $(BJ^-)_{-p-2h,k+h,l}$ in Proposition 14.4. Then we have

$$\begin{aligned}
&-2(\sqrt{c_1}a_1)(\sqrt{-c_2}a_2)\{-2(h+1)\}(h+k+1)\, b_{-p-2h-2,k+h+1,l}(a) \\
&= (r-k-h)\Big[2c_2a_2^2\partial_1 - 2c_2a_2^2(r+2+h+\frac{p-u}{2}) \\
&\qquad +2c_1a_1^2\partial_2 - 2c_1a_1^2(r+2+h+\frac{p+u}{2})\Big] \\
&\qquad \times \binom{k+h}{h}^{-1} \binom{r-k}{h} \mathcal{R}_h^{(p,u)}(\sqrt{-c_2}a_2, \sqrt{c_1}a_1)\, b_{-p,k,l}(a).
\end{aligned}$$

Therefore we have,

$$\begin{aligned}
&4(h+1)(\sqrt{c_1}a_1)(\sqrt{-c_2}a_2)\, b_{-p-2h-2,k+h+1,l}(a) \\
&= \binom{k+h+1}{h+1}^{-1} \binom{r-k}{h+1} \Big[2c_2a_2^2\partial_1 - 2c_2a_2^2(r+2+h+\frac{p-u}{2}) \\
&\quad +2c_1a_1^2\partial_2 - 2c_1a_1^2(r+2+h+\frac{p+u}{2})\Big] \\
&\quad \times \mathcal{R}_h^{(p,u)}(\sqrt{-c_2}a_2, \sqrt{c_1}a_1)\, b_{-p,k,l}(a).
\end{aligned}$$

And use the equality,

$$\begin{aligned}
&\Big[2c_2a_2^2\partial_1 - 2c_2a_2^2(r+2+h+\frac{p-u}{2}) \\
&\quad +2c_1a_1^2\partial_2 - 2c_1a_1^2(r+2+h+\frac{p+u}{2})\Big] \\
&\qquad \times \mathcal{R}_h^{(p,u)}(\sqrt{-c_2}a_2, \sqrt{c_1}a_1)\, b_{-p,k,l}(a) \\
&= \Big\{(2c_2a_2^2\partial_1 + 2c_1a_1^2\partial_2)\mathcal{R}_h^{(p,u)}(\sqrt{-c_2}a_2, \sqrt{c_1}a_1)\Big\} \cdot b_{-p,k,l}(a) \\
&\quad +\mathcal{R}_h^{(p,u)}(\sqrt{-c_2}a_2, \sqrt{c_1}a_1)\}\Big[2c_2a_2^2\partial_1 - 2c_2a_2^2(r+2+h-\frac{p-u}{2}) \\
&\quad +2c_1a_1^2\partial_2 - 2c_1a_1^2(r+2+h-\frac{p+u}{2}) \\
&\qquad -2c_2a_2^2(h+p-u) - 2c_1a_1^2(h+p+u)\Big]\, b_{-p,k,l}(a).
\end{aligned}$$

Thus we get

$$
\begin{aligned}
&-2(h+1)\cdot\binom{k+h+1}{h+1}\binom{r-k}{h+1}^{-1} b_{-p-2h-2,k+h+1,l}(a)\\
&=\Big\{\Big[(\sqrt{-c_2}a_2)(\sqrt{c_1}a_1)^{-1}\partial_1-(\sqrt{-c_2}a_2)^{-1}(\sqrt{c_1}a_1)\partial_2\\
&\qquad -(\sqrt{-c_2}a_2)(\sqrt{c_1}a_1)^{-1}(h+p-u)+(\sqrt{-c_2}a_2)^{-1}(\sqrt{c_1}a_1)(h+p+u)\Big]\\
&\qquad \times\mathcal{R}_h^{(p,u)}(\sqrt{-c_2}a_2,\sqrt{c_1}a_1)\Big\}\cdot b_{-p,k,l}(a)\\
&=-2(h+1)\mathcal{R}_{h+1}^{(p,u)}(\sqrt{-c_2}a_2,\sqrt{c_1}a_1)\, b_{-p,k,l}(a).
\end{aligned}
$$

In the last calculation, we used Lemma 14.12 ($x=\sqrt{-c_2}a_2$, $y=\sqrt{c_1}a_1$, $\alpha=p$ and $\beta=u$). □

14.4. Main theorem for indefinite characters.

THEOREM 14.14. *Let* $\pi=\pi_\Lambda^*$, $\Lambda=[r-1,s+1;u]\in\Xi_{\mathrm{III}}$, $(r-s-2>|u|)$, *an indefinite character* $\xi=\mathrm{diag}(c_1,c_2)$ *of* N_S *with* $c_1>0$ *and* $c_2<0$, *and* χ *be an irreducible unitary representation of* $SU(\xi)$ $(\simeq SU(1,1))$. *Set* $\eta=\chi\otimes\xi$.

(a) *If* χ *is a unitary principal series representation, a complementary series representation or the trivial representation then*

$$\dim_{\mathbb{C}}\mathrm{Hom}_{(\mathfrak{g},K)}(\pi_\Lambda^*,C_\eta^\infty(R\backslash G))=0. \tag{184}$$

(b-1) *If* χ *is (limits of) the discrete series representation* $D_p^\pm$ *with Blattner parameter* $\pm p$ *under the condition that* $p<r-s$ *or* $p>r+s$, *then*

$$\dim_{\mathbb{C}}\mathrm{Hom}_{(\mathfrak{g},K)}(\pi_\Lambda^*,C_\eta^\infty(R\backslash G))=0. \tag{185}$$

(b-2) *If* χ *is the discrete series representation* D_p^+ *with Blattner parameter* p *under the condition that* $r-s\le p\le r+s$ *and* $p\equiv r+s$ (mod 2), *then*

$$\dim_{\mathbb{C}}\mathrm{Hom}_{(\mathfrak{g},K)}(\pi_\Lambda^*,C_\eta^\infty(R\backslash G))=1. \tag{186}$$

In this case, we have moreover,

$$\dim_{\mathbb{C}}\mathrm{Hom}_{(\mathfrak{g},K)}(\pi_\Lambda^*,C_\eta^\infty(R\backslash G)^{\mathrm{mod}})=0. \tag{187}$$

More precisely, if we write $p=r-s+2q$ *by using an integer* $0\le q\le s$, *then the generalized Whittaker functions* $\Phi_{\pi,\tau}(a)$ *for* $\pi=\pi_\Lambda^*$ *with minimal* K*-type* τ^*, $\tau=[r,s;u]$ *are described as follows.*
Let $\Phi_{\pi,\tau}(a)=\sum_{j\in p+2\mathbb{Z}_{\ge0}}\sum_{\substack{0\le k\le r\\0\le l\le s}} b_{jkl}(a)(u_j\otimes f_{kl})\in C_{M_0}^\infty(A;V_p\otimes W_\tau)$,

then $\Phi_{\pi,\tau}(a)$ *can be written and up to constant uniquely determined as*

$$\Phi_{\pi,\tau}(a) = \sum_{w=0}^{s-q} \sum_{h=0}^{s-q-w} b_{p+2h,s-q-w-h,w}(a)(u_{p+2h} \otimes f_{s-q-w-h,w}),$$

and the coefficient functions are given by

$$\begin{aligned} & b_{p+2h,s-q-w-h,w}(a) \\ & = \binom{r}{s-q-w-h}\binom{r}{s-q}^{-1}(w!)^{-1} \\ & \times \mathcal{R}_h^{(p,u)}(\sqrt{c_1}a_1, \sqrt{-c_2}a_2) \cdot \mathcal{F}_w^+(-4\pi c_1 a_1^2 + 4\pi c_2 a_2^2) \\ & \times (\sqrt{c_1}a_1)^{r+2-b_2-q}(\sqrt{-c_2}a_2)^{r+2+b_1-q} \\ & \times (c_1 a_1^2 - c_2 a_2^2)^{-(s-q+1)} e^{2\pi(c_1 a_1^2 - c_2 a_2^2)}. \end{aligned} \tag{188}$$

Therefore, all $b_{jkl}(a)$ *are exponentially increasing.*

(b-3) *If* χ *is the discrete series representation* D_p^- *with Blattner parameter* $-p$ *under the condition that* $r - s \leq p \leq r + s$ *and* $p \equiv r + s$ (mod 2), *then*

$$\dim_{\mathbb{C}} \mathrm{Hom}_{(\mathfrak{g},K)}(\pi_\Lambda^*, C_\eta^\infty(R\backslash G)) = 1. \tag{189}$$

In this case, we have moreover,

$$\dim_{\mathbb{C}} \mathrm{Hom}_{(\mathfrak{g},K)}(\pi_\Lambda^*, C_\eta^\infty(R\backslash G)^{\mathrm{mod}}) = 1. \tag{190}$$

More precisely, if we write $p = r - s + 2q$ *by using an integer* $0 \leq q \leq s$, *then the generalized Whittaker functions* $\Phi_{\pi,\tau}(a)$ *for the representation* $\pi = \pi_\Lambda^*$ *with minimal* K*-type* τ^*, $\tau = [r, s; u]$ *are described as follows. Let* $\Phi_{\pi,\tau}(a) = \sum_{j\in -p+2\mathbb{Z}_{\leq 0}} \sum_{\substack{0\leq k\leq r \\ 0\leq l\leq s}} b_{jkl}(a)(u_j \otimes f_{kl}) \in C_{M_0}^\infty(A; \bar{V}_p \otimes W_\tau)$, *then* $\Phi_{\pi,\tau}(a)$ *can be written and up to constant uniquely determined as*

$$\Phi_{\pi,\tau}(a) = \sum_{w=0}^{s-q} \sum_{h=0}^{s-q-w} b_{-p-2h,r-s+q+w+h,s-w}(a)(u_{-p-2h} \otimes f_{r-s+q+w+h,s-w}),$$

and the coefficient functions are given by

$$\begin{aligned} & b_{-p-2h,r-s+q+w+h,s-w}(a) \\ & = \binom{r}{s-q-w-h}\binom{r}{s-q}^{-1}(w!)^{-1} \\ & \times \mathcal{R}_h^{(p,u)}(\sqrt{-c_2}a_2, \sqrt{c_1}a_1) \cdot \mathcal{F}_w^+(4\pi c_1 a_1^2 - 4\pi c_2 a_2^2) \\ & \times (\sqrt{c_1}a_1)^{r+2+b_1-q}(\sqrt{-c_2}a_2)^{r+2-b_2-q} \\ & \times (c_1 a_1^2 - c_2 a_2^2)^{-(s-q+1)} e^{-2\pi(c_1 a_1^2 - c_2 a_2^2)}. \end{aligned} \tag{191}$$

Actually all $b_{jkl}(a)$ are rapidly decreasing.
Here,

$$\mathcal{F}_w^+(x) = \sum_{i=0}^{w} \binom{w}{i} \frac{(s-w+i)!}{(s-w)!} x^{w-i},$$

and

$$\mathcal{R}_h^{(p,u)}(x,y) = \sum_{i=0}^{h} (-1)^i \binom{\frac{p-u}{2}+h-i-1}{h-i} \binom{\frac{p+u}{2}+i-1}{i} x^{h-2i} y^{-h+2i}.$$

Remark. The coefficient functions $b_{p+2h,s-q-w-h,w}(a)$ in **(b-2)** are identical with
$b_{-p-2h,r-s+q+w+h,s-w}(a)$ in **(b-3)** when we interchange the variables $(\sqrt{c_1}a_1, \sqrt{-c_2}a_2)$ into $(\sqrt{-c_2}a_2, \sqrt{c_1}a_1)$.

PROOF. Propositions 14.3 and 14.7 imply the case **(a)**. The case **(b-1)** was proved in Proposition 14.6. Lemma 14.8, Propositions 14.10 and 14.13 imply the case **(b-2)** and **(b-3)**. Actually these $b_{jkl}(a)$ satisfy partial differential equations $(B_2^+ : 1)_{jkl}, \ldots, (B_2^- : 2)_{jkl}$ in Proposition 14.2, which is equivalent to that of Proposition 14.4. Note that

$$\binom{r}{s-q-w-h}\binom{r}{s-q}^{-1}$$
$$= \binom{r-s+q+w+h}{h}^{-1} \binom{r-s+q+w}{w}^{-1} \binom{s-q-w}{h} \binom{s-q}{w}$$

and $\mathcal{F}_w^+(-x) = (-1)^w \mathcal{F}_w^-(x)$ in the explicit formula in **(b-2)**. □

Bibliography

[DG1] A. Debiard and B. Gaveau, *Répresentation intégrale de certaines séries de fonctions sphériques d'un systèm de racines BC.* J. Funct. Anal. **96** (1991), 256-296.

[DG2] A. Debiard and B. Gaveau, *Integral formulas for the spherical polynomials of a root system of type BC_2.* J. Funct. Anal. **119** (1994), 401-454.

[Ha1] T. Hayata, *Differential equations for principal series Whittaker functions on $SU(2,2)$.* Indag. Mathem., N.S., **8** (4),(1997), 493-528.

[Ha2] T. Hayata, *Whittaker functions of generalized principal series on $SU(2,2)$.* J. Math. Kyoto Univ. **37-3** (1997), 531-546.

[HaO] T. Hayata and T. Oda, *An explicit integral representation of Whittaker functions for the representations of the discrete series —the case of $SU(2,2)$—* . J. Math. Kyoto Univ. **37-3** (1997), 519-530.

[Ho] A. Hori, *Andrianov's L-functions associated to Siegel wave forms of degree two.* Math. Ann. **303** (1995), 195-226.

[Ii] M. Iida, *Spherical functions of the principal series representations of $Sp(2,\mathbf{R})$ as hypergeometric functions of C_2-type.* Publ. RIMS, Kyoto Univ. **32** (1996), 689-727.

[Kn] A. W. Knapp, *Representation theory of semisimple groups, — an overview based on examples—*, Princeton Mathematical Series **36** (1986), 1-773.

[Koe] M. Koecher, *Zur Theorie der Modulformen n-ten Grades. I.* Math. Z. **59** (1959), 399-416.

[Kos] B. Kostant, *On Whittaker vectors and representation theory.* Invent. Math. **48** (1978), 101-184.

[Mi] T. Miyazaki, *The generalized Whittaker functions for $Sp(2,\mathbf{R})$ and gamma factor of the Andrianov L-function.* preprint, RIMS-1053. (1995), 1-74.

[Ni] S. Niwa, *On generalized Whittaker functions on Siegel's upper half space of degree 2.* Nagoya Math. J. **121** (1991), 171-184.

[Vog] D. Vogan Jr., *Gelfand-Kirillov dimension for Harish-Chandra modules.* Invent. Math. **49** (1978), 75-98.

[Wal] N. Wallach, *Asymptotic expansions of generalized matrix entries of representations of real reductive groups.* Lie Group Representations I, Lecture Notes in Math. **1024**, Springer-Verlag (1984), 287-369.

[WW] E. T. Whittaker and G. N. Watson, *A course of modern analysis.* Cambridge Univ. Press, 4ed., (1965), 1-608.

[Ya1] H. Yamashita, *Finite multiplicity theorems for induced representations of semisimple Lie groups I, II.* J. Math. Kyoto Univ. **28** (1988), 173-211, 383-444.

[Ya2] H. Yamashita, *Embeddings of discrete series into induced representations of semisimple Lie groups, I: General theory and the case of $SU(2,2)$*. Japan J. Math. (N.S.), **16** (1990), 31-95.

DEPARTMENT OF MATHEMATICS, FACULTY OF SCIENCE, SAITAMA UNIVERSITY, URAWA, SAITAMA, 338-8570, JAPAN
E-mail: ygon@rimath.saitama-u.ac.jp

Editorial Information

To be published in the *Memoirs*, a paper must be correct, new, nontrivial, and significant. Further, it must be well written and of interest to a substantial number of mathematicians. Piecemeal results, such as an inconclusive step toward an unproved major theorem or a minor variation on a known result, are in general not acceptable for publication. Papers appearing in *Memoirs* are generally longer than those appearing in *Transactions*, which shares the same editorial committee.

As of September 30, 2001, the backlog for this journal was approximately 7 volumes. This estimate is the result of dividing the number of manuscripts for this journal in the Providence office that have not yet gone to the printer on the above date by the average number of monographs per volume over the previous twelve months, reduced by the number of volumes published in four months (the time necessary for preparing a volume for the printer). (There are 6 volumes per year, each containing at least 4 numbers.)

A Consent to Publish and Copyright Agreement is required before a paper will be published in the *Memoirs*. After a paper is accepted for publication, the Providence office will send a Consent to Publish and Copyright Agreement to all authors of the paper. By submitting a paper to the *Memoirs*, authors certify that the results have not been submitted to nor are they under consideration for publication by another journal, conference proceedings, or similar publication.

Information for Authors

Memoirs are printed from camera copy fully prepared by the author. This means that the finished book will look exactly like the copy submitted.

The paper must contain a *descriptive title* and an *abstract* that summarizes the article in language suitable for workers in the general field (algebra, analysis, etc.). The *descriptive title* should be short, but informative; useless or vague phrases such as "some remarks about" or "concerning" should be avoided. The *abstract* should be at least one complete sentence, and at most 300 words. Included with the footnotes to the paper should be the 2000 *Mathematics Subject Classification* representing the primary and secondary subjects of the article. The classifications are accessible from `www.ams.org/msc/`. The list of classifications is also available in print starting with the 1999 annual index of *Mathematical Reviews*. The Mathematics Subject Classification footnote may be followed by a list of *key words and phrases* describing the subject matter of the article and taken from it. Journal abbreviations used in bibliographies are listed in the latest *Mathematical Reviews* annual index. The series abbreviations are also accessible from `www.ams.org/publications/`. To help in preparing and verifying references, the AMS offers MR Lookup, a Reference Tool for Linking, at `www.ams.org/mrlookup/`. When the manuscript is submitted, authors should supply the editor with electronic addresses if available. These will be printed after the postal address at the end of the article.

Electronically prepared manuscripts. The AMS encourages electronically prepared manuscripts, with a strong preference for AMS-LaTeX. To this end, the Society has prepared AMS-LaTeX author packages for each AMS publication. Author packages include instructions for preparing electronic manuscripts, the *AMS Author Handbook*, samples, and a style file that generates the particular design specifications of that publication series. Though AMS-LaTeX is the highly preferred format of TeX, author packages are also available in AMS-TeX.

Authors may retrieve an author package from e-MATH starting from www.ams.org/tex/ or via FTP to ftp.ams.org (login as anonymous, enter username as password, and type cd pub/author-info). The *AMS Author Handbook* and the *Instruction Manual* are available in PDF format following the author packages link from www.ams.org/tex/. The author package can be obtained free of charge by sending email to pub@ams.org (Internet) or from the Publication Division, American Mathematical Society, P.O. Box 6248, Providence, RI 02940-6248. When requesting an author package, please specify AMS-LaTeX or AMS-TeX, Macintosh or IBM (3.5) format, and the publication in which your paper will appear. Please be sure to include your complete mailing address.

Sending electronic files. After acceptance, the source file(s) should be sent to the Providence office (this includes any TeX source file, any graphics files, and the DVI or PostScript file).

Before sending the source file, be sure you have proofread your paper carefully. The files you send must be the EXACT files used to generate the proof copy that was accepted for publication. For all publications, authors are required to send a printed copy of their paper, which exactly matches the copy approved for publication, along with any graphics that will appear in the paper.

TeX files may be submitted by email, FTP, or on diskette. The DVI file(s) and PostScript files should be submitted only by FTP or on diskette unless they are encoded properly to submit through email. (DVI files are binary and PostScript files tend to be very large.)

Electronically prepared manuscripts can be sent via email to pub-submit@ams.org (Internet). The subject line of the message should include the publication code to identify it as a Memoir. TeX source files, DVI files, and PostScript files can be transferred over the Internet by FTP to the Internet node e-math.ams.org (130.44.1.100).

Electronic graphics. Comprehensive instructions on preparing graphics are available at www.ams.org/jourhtml/graphics.html. A few of the major requirements are given here.

Submit files for graphics as EPS (Encapsulated PostScript) files. This includes graphics originated via a graphics application as well as scanned photographs or other computer-generated images. If this is not possible, TIFF files are acceptable as long as they can be opened in Adobe Photoshop or Illustrator. No matter what method was used to produce the graphic, it is necessary to provide a paper copy to the AMS.

Authors using graphics packages for the creation of electronic art should also avoid the use of any lines thinner than 0.5 points in width. Many graphics packages allow the user to specify a "hairline" for a very thin line. Hairlines often look acceptable when proofed on a typical laser printer. However, when produced on a high-resolution laser imagesetter, hairlines become nearly invisible and will be lost entirely in the final printing process.

Screens should be set to values between 15% and 85%. Screens which fall outside of this range are too light or too dark to print correctly. Variations of screens within a graphic should be no less than 10%.

Inquiries. Any inquiries concerning a paper that has been accepted for publication should be sent directly to the Electronic Prepress Department, American Mathematical Society, P. O. Box 6248, Providence, RI 02940-6248.

Editors

Selected Titles in This Series

(*Continued from the front of this publication*)

For a complete list of titles in this series, visit the
AMS Bookstore at **www.ams.org/bookstore/**.